AF837045

Gas Plasma Sterilization in Microbiology

Theory, Applications, Pitfalls and New Perspectives

Edited by

Hideharu Shintani

Faculty of Science and Engineering
Chuo University
Tokyo
Japan

and

Akikazu Sakudo

Laboratory of Biometabolic Chemistry
School of Health Sciences
University of Ryukyus
Okinawa
Japan

Caister Academic Press

Caister Academic Press
Norfolk, UK

www.caister.com

British Library Cataloguing-in-Publication Data
A catalogue record for this book is available from the British Library

ISBN: 978-1-910190-25-8 (paperback)
ISBN: 978-1-910190-26-5 (ebook)

Description or mention of instrumentation, software, or other products in this book does not imply endorsement by the author or publisher. The author and publisher do not assume responsibility for the validity of any products or procedures mentioned or described in this book or for the consequences of their use.

Cover design adapted from Figures 10.1 and 10.2.

Contents

Contributors

Pervin Başaran Akocak
Department of Food Engineering
Suleyman Demirel University
Isparta
Turkey

pervinbasaran@sdu.edu.tr

Takeshi Azuma
Department of Gastroenterology
Kobe University
Kobe
Japan

azumat@med.kobe-u.ac.jp

Eiki Hotta
Department of Energy Sciences
Tokyo Institute of Technology
Yokohama
Japan

ehotta@es.titech.ac.jp

Yuichiro Imanishi
NGK Insulators, Ltd
Nagoya
Japan

yimanishi@ngk.co.jp

Tatsuya Misawa
Department of Electrical and Electronic Engineering
Faculty of Science and Engineering
Saga University
Saga
Japan

misawa@cc.saga-u.ac.jp

Tomoko Mishima
Research Institute of Environment, Agriculture and Fisheries, Osaka Prefecture
Osaka
Japan

mishimat@mbox.kannousuiken-osaka.or.jp

Hidekazu Miyahara
Department of Energy Sciences
Tokyo Institute of Technology
Yokohama
Japan

miya@plasma.es.titech.ac.jp

Terumi Nishioka
Research Institute of Environment, Agriculture and Fisheries, Osaka Prefecture
Osaka
Japan

nishioka@mbox.kannousuiken-osaka.or.jp

Akitoshi Okino
Department of Energy Sciences
Tokyo Institute of Technology
Yokohama
Japan

aokino@es.titech.ac.jp

Gyungsoon Park
Plasma Bioscience Research Center
Kwangwoon University
Seoul
Republic of Korea

gyungp@kw.ac.kr

Akikazu Sakudo
Laboratory of Biometabolic Chemistry
School of Health Sciences
University of the Ryukyus
Okinawa
Japan

sakudo@med.u-ryukyu.ac.jp

Naohiro Shimizu
NGK Insulators, Ltd
Nagoya
Japan

shimizu@plasma.engg.nagoya-u.ac.jp

Hideharu Shintani
Faculty of Science and Engineering
Chuo University
Tokyo
Japan

hshintani@jcom.zaq.ne.jp

Toshihiro Takamatsu
Department of Gastroenterology
Kobe University
Kobe; and
Department of Energy Sciences
Tokyo Institute of Technology
Yokohama
Japan

tplasma@med.kobe-u.ac.jp

Yoichi Toyokawa
Laboratory of Biometabolic Chemistry
School of Health Sciences
University of the Ryukyus
Okinawa
Japan

yoichitoyokawa8192@gmail.com

Takuya Uyama
Department of Energy Sciences
Tokyo Institute of Technology
Yokohama
Japan

uyama.t@m.titech.ac.jp

Yoshihito Yagyu
Department of Electrical and Electric Engineering
National Institute of Technology
Sasebo College
Nagasaki
Japan

yyagyu@sasebo.ac.jp

Preface

Gas plasma is the fourth state of matter, alongside solid, liquid and gas. There are many naturally occurring events and man-made products related to gas plasma including aurora and thunderstorms, and high-intensity discharge (HID) headlamp bulbs, oxonizers, semiconductors and solar battery panels. As a result, gas plasma technology is increasingly important in our life.

Among the various technologies, particular attention should be paid to the use of gas plasma in sterilization and disinfection. Gas plasma treatment has helped to minimize the contamination of medical instruments with infectious pathogens and toxins and, thus, the prevention of hospital-acquired infection.

The purpose of this book is to bring together information on the current status and future prospects of the state-of-art physical technique of gas plasma sterilization. The chapters cover basic information on this method of sterilization, applications of gas plasma technology to the inactivation of toxins and pathogens, possible mechanisms of gas plasma sterilization, and verification and validation of the sterilization efficiency of gas plasma, as well as discussing the challenges, limitations, and advantages of gas plasma sterilization, as well as future research perspectives.

This book will provide a standard reference and indispensable roadmap of gas plasma sterilization for students, engineers, and laboratory scientists. I hope that readers will enjoy this book, obtain useful information for their own research, and be inspired by new ideas for future research on gas plasma sterilization.

Akikazu Sakudo

Acknowledgements

I sincerely thank my wife, Miharu Shintani, who supported me during the preparation of this book.

Hideharu Shintani

I am pleased to have the honour of compiling this book together with Dr Shintani and to have been given the opportunity to work with such eminent scientists as the chapter contributors, whose combined effort have made this book possible. In addition, we would like to acknowledge the grant-in-aids, especially grant-in-aid for science and technology research promotion programme for agriculture, forestry, fisheries and food industry, which supports the work published in this book. We also thank the publishers for granting permission to use previously published figures that are included in this book. Finally, we wish to thank Annette Griffin and the other editorial staff at Caister Academic Press for their professionalism and dedication.

Akikazu Sakudo

Introduction

1

Hideharu Shintani

Abstract

Gas plasma sterilization offers enormous potential as a broad-spectrum antimicrobial procedure. In this chapter we explain the types of errors that result from the 'understanding gap' between the engineering researchers who are developing the gas plasma sterilization technology and the microbiologists who aim to fine-tune it for their needs. Future initiatives to exploit this powerful technology would benefit from adopting multidisciplinary approaches, involving close collaboration between microbiologists, chemists and engineering researchers.

Introduction

Gas plasma sterilization is currently of interest mainly to engineering researchers; however, in many cases their publications contain significant errors because they do not have a background/understanding of microbiology and sterilization. As a result, several misconceptions about the use and efficacy of the gas plasma sterilization process have been published. As a result relatively few microbiological and chemical researchers are involved in gas plasma sterilization research. In 2011, Sakudo and Shintani published a book in which they noted that many current publications on similar topics contain several mistakes in their discussion and interpretation of microbiology and sterilization. For that reason the authors included a chapter entitled 'Several Points to Consider When Conducting Plasma Experiments'. However, no revisions or corrections have been published in the literature since then. Therefore, the goal of this new gas plasma sterilization book is to summarize new up-to-date information and critically evaluate the current status of the technology.

Common data interpretation errors

For example, one misinterpretation of the engineering studies is in regard to the meaning of the required '6 log reduction' for sterility assurance. A 6 log reduction does not mean zero, as $10^0 = 1$, and the possibility of survival at a sterility assurance level (SAL) of 10^0 is 63%. A 6 log reduction is the requirement for the absolute bioburden method in ISO 14161, which addresses biological indicator (BI) users only. The bioburden is the type and number of viable microorganisms in or on the product, and in sterilization validation the real target of sterilization is not the BI, but rather the bioburden. Bioburdens of 10^6 CFU (colony-forming units)/carrier do not exist in real-life situations. For example, according to the absolute bioburden method in ISO 14161, an initial population of 10^0 CFU/carrier level (i.e. a few CFU) that undergoes a 6 log reduction attains a SAL of 10^{-6}; 10^{-6} is the closest number to zero, which was defined from the stochastics in ISO 11137-1.

To achieve a 6 log or 12 log reduction, the survivor curve must be a straight line. BI manufacturers are required to use an initial population of 10^6 CFU/carrier, so reduction by 6 logs to a SAL of 10^{-6} represents a 12 log reduction: this is required in ISO 11138-1 as an overkill method for sterilization validation. An initial population of 10^6 CFU/carrier to a SAL of 10^0 is not recognized as a 6 log reduction in ISO 14161. Since ISO 11138-1 is for BI manufacturers and ISO 14161 is for BI users, researchers, as BI users, must obey the ISO 14161 requirements for sterilization validation. The main difference between ISO 11138-1 (BI manufacturer) and ISO 14161 (BI user) is that in ISO 14161, the overkill method is described together with another method, but in ISO 11138-1, only the overkill method is described. Detailed information about ISO requirements will be discussed in Chapter 13.

Stacking is often mistaken as clumping in the BI (Fig. 1.1), and the presence of clumping in the BI (biological indicator) is a serious problem. In order to attain a straight survivor curve from an initial population of 10^6 CFU/carrier to a SAL of 10^{-6} (a 12 log reduction), the BI should be free from any clumping (Fig. 1.2). A 12 log reduction is required for BI manufacturers in ISO 11138-1, but it is not always required of the BI user in ISO 14161. To attain a 12 log reduction, it is necessary to avoid clumping in the BI; otherwise a curved (tailing) survivor curve is obtained (Fig. 1.3A). In this case even a SAL of 10^0 cannot be attained, indicating that sterilization validation failed. Official documentation of sterilization validation is required, and relevant authorities conduct inspections to confirm sterilization validation. The straight line survivor curves shown in Fig. 1.3B and Fig. 1.4 were obtained using the BI shown in Fig. 1.2, which was free from clumping. These curves were obtained with 10^6 CFU/carrier and a SAL of 10^{-6}, which represents a full 12 log reduction (Figs. 1.3B and 1.4). ISO 11138-1 requires that the coefficient correlation of the survivor curve must be greater than 0.8 (ISO 11138-1, Normative Annex B).

It is quite important to note that the *D* value (decimal reduction value, i.e., the time or dose required to decrease by 1 log) is only one per one microorganism. Often the tailing phenomenon (Fig. 1.3A) can be explained by a difference in the kinetics of killing. Therefore, *D* values are calculated for each kinetic curve, indicating that more than one *D* value

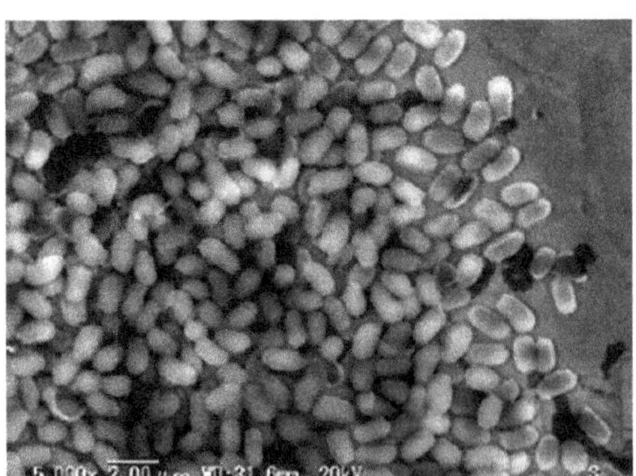

Figure 1.1 SEM observation of pile of clumping. Reproduced with permission from Shintani *et al.* (2010).

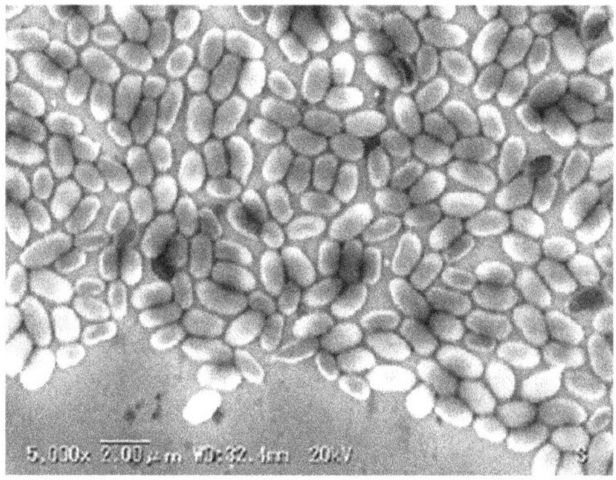

Figure 1.2 SEM observation free from clumping. Reproduced with permission from Shintani *et al.* (2010).

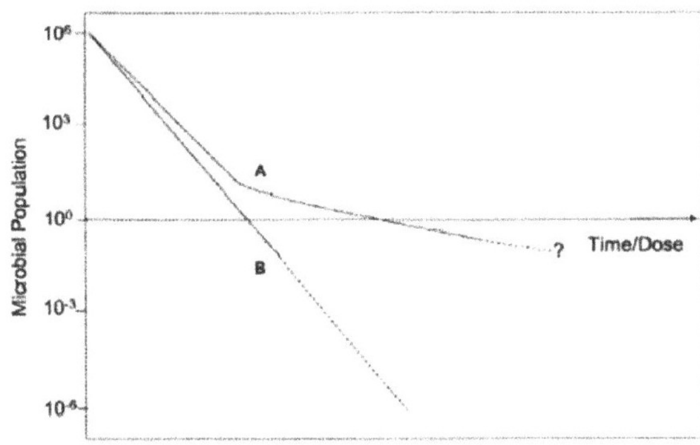

Figure 1.3 Straight line and curved tailing survivor curve. Reproduced from McDonnell, G.E. (2007) Antisepsis, Disinfection, and Sterilization, Types, Action, and Resistance, ASM Press, NW, Washington DC, with permission from ASM Press.

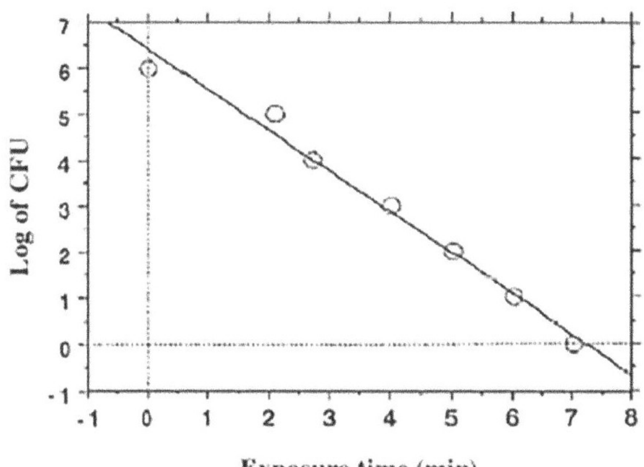

Figure 1.4 Straight line of survivor curve using BI from Fig. 1.2. When using a clump-free BI (Fig. 1.2), a straight line can be experimentally confirmed from an initial population of 10^6 CFU/carrier to a SAL 10^{-1}. A 6 log reduction takes 7 min, indicating that the *D* value is 1.2 min. Reproduced with permission from Shintani *et al.* (2010).

was present per microorganism, which is a serious mistake. Why does clumping cause tailing? The penetration depth of gas plasma is quite shallow (~10–20 nm) except for oxygen gas plasma, so when spores are in multiple layers (clumping), only those at the surface of the first layer are immediately killed. As a result, killing of the second or third layer is delayed by the interference of the first layer, and this delay causes the observed tailing of the curve (Fig. 1.3A). Tailed survivor curves are not the true survivor curves. The BI *Geobacillus stearothermophilus* ATCC 7953 is the standard endospore former used in gas plasma sterilization. The average size of *G. stearothermophilus* spores is 1 μm × 3 μm (rod) (Shintani *et al.*, 2007), so if the BI has clumps, those spores below the surface layer will not be killed because the penetration depth of gas plasma sterilization is ~ 10–20 nm.

Advantage of gas plasma sterilization

Because gas plasma penetration is so shallow (~10–20 nm), the bioburden is killed and the product itself is not damaged. To kill the bioburden without deterioration of the material being sterilized is called simultaneous attainment of material/functional compatibility and a SAL of 10^{-6}. The bioburden is scattered on the surface of the product and 10–20 CFU/500 l (Shintani *et al.*, 2004, 2006) without clumping is the real estimated bioburden. This level of bioburden is scattered over a larger area of the product than the BI, so no clumping is observed, and thus no tailing phenomenon is observed for the bioburden following ISO 14161 (absolute bioburden method).

Conclusion

In order to avoid further misinterpretation by the engineering researchers, we, as microbiologists and chemists, need to contribute by conveying appropriate information and correct mechanisms to the engineers. Recently obtained sterilization mechanisms using spores are presented in this book. Additional useful information is also presented, including ISO descriptions needed for future validation studies.

Note that the technology utilized in the Sterrad® sterilizer (J&J Company) is beyond the scope of this book so is not included.

References

Sakudo, A., and Shintani, H. (2011). Sterilization and Disinfection by Plasma: Sterilization Mechanisms, Biological and Medical Applications. NOVA Science Publishers, Hauppauge, NY, USA.

Shintani, H., Taniai, E., Miki, A., Kurosu, S., and Hayashi, F. (2004). Comparison of the collecting efficiency of microbiological air samplers. J. Hosp. Infect. *56*, 42–48.

Shintani, H., Kurosu, S., Miki, A., Hayashi, F., and Kato, S. (2006). Sterilization efficiency of the photocatalysts against environmental microorganisms in a health care facility. Biocontrol Sci. *11*, 17–26.

Shintani, H., Shimizu, N., Imanishi, Y., Sekiya, Y., Tamazawa, K., Taniguchi, A., and Kido, N. (2007). Inactivation of microorganisms and endotoxins by low temperature nitrogen gas plasma exposure. Biocontrol Sci. *12*, 131–143.

Shintani, H., Sakudo, A., Burke, P., and McDonnell, G. (2010). Gas plasma sterilization of microorganisms and mechanisms of action. Exp. Ther. Med. *1*, 731–738.

Theoretical Background and Mode of Action of Gas Plasma Sterilization

2

Hideharu Shintani

Abstract

In this chapter theoretical background and potential mode of action of gas plasma sterilization are described. Nowadays, gas plasma sterilization is widely utilized for sterilization of spores. Because spores are more tolerant than vegetative bacterial cells, bacterial endospores are used as the biological indicator (BI). *Geobacillus stearothermophilus* ATCC 7953 is generally used as the BI for gas plasma sterilization. Gas plasma sterilization penetration is quite shallow, i.e. ~10–20 nm from the surface, so it is easy to achieve a sterility assurance level (SAL) of 10^{-6} and material/functional compatibility compared with alternative sterilization procedures. Simultaneous achievement of a SAL of 10^{-6} and material/functional compatibility will be discussed in detail in Chapter 3.

Introduction

Ninety-nine per cent of the material in the universe is in the plasma state. Physicists call plasma the fourth state of matter, after solid, liquid and gaseous states. Basically, plasma is composed of gas molecules that have been dissociated by an input of energy.

Low-temperature gas plasmas are generated when certain gases are stimulated at atmospheric pressure (AP) or relatively low pressure (LP) with pulsed energy, radio-frequency waves or microwave energy. The plasmas of several different gases such as argon, helium, oxygen, nitrogen or their mixtures have sporicidal activity (Shintani and MacDonnell, 2011; Lassen *et al.*, 2003; Rossi and Kylian, 2012). Many modern medical devices are thermo- and hydro-sensitive. Given the drawbacks and limitations of other low-temperature sterilization procedures, low temperature gas plasma sterilization could represent a useful alternative. Model equipment used for gas plasma sterilization is shown in Fig. 2.1 (Shintani *et al.*, 2007).

The first practical application of gas plasma sterilization was developed in 1972 (Ascham and Menashi, 1972). Since then several kinds of gas plasma applications have been designed to sterilize the bioburden of various products, and many researchers have studied the subject as well as the mechanism of gas plasma sterilization. The bioburden defines the type and number of viable microorganisms in/on a product (ISO 14161, 11138-1).

Low-temperature gas plasmas, used for surface modification and organic cleaning (Fig. 2.2), are ionized gases generated at pressures between 0.1 and 2 torr. These types of plasmas work within a vacuum chamber from which atmospheric gases have been evacuated, typically below 0.1 torr. Low pressure allows for a relatively long free path of accelerated radicals and metastables. Neutral particles such as radicals and metastables can be produced

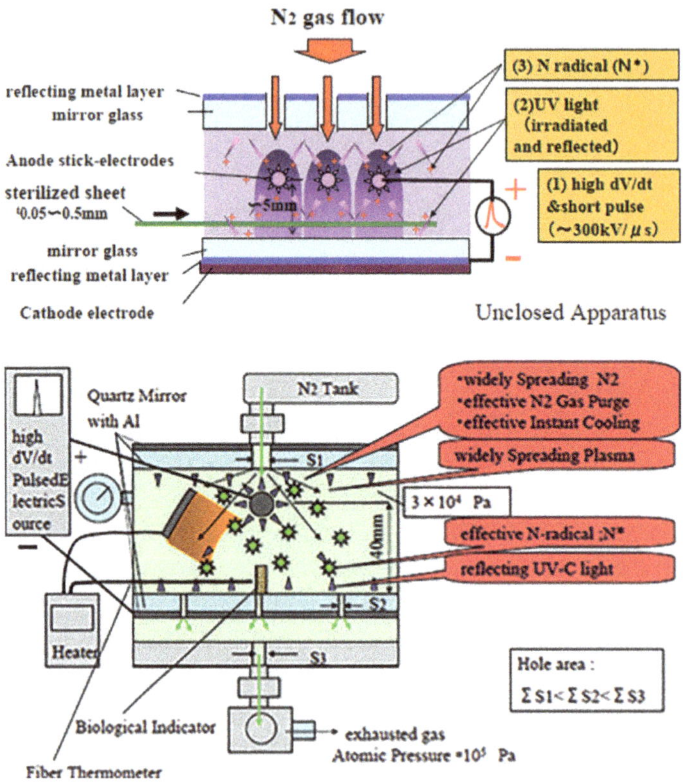

Figure 2.1 Design of gas plasma sterilizer. Above: atmospheric pressure gas plasma sterilizer (AP). Below: Relatively low (1/2 to 1/3*P*) pressure gas plasma sterilizer. The gap between anode and cathode is 4 mm for AP and 40 to 150 mm for LP. Modified from Shintani *et al*. (2007).

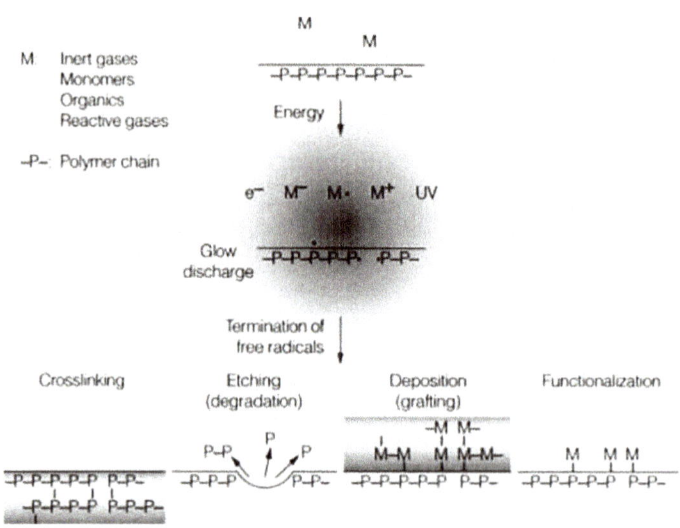

Figure 2.2 Surface modification by gas plasma exposure. Cited from URL of http://www.astp.com/plasma/pl_examples.html with permission from AST Products Inc.

at or near ambient temperatures, and undergo relatively few collisions with molecules at this temperature. Although both radicals and metastables are neutral compounds, their flight distances differ significantly. Whereas the flight distance of a radical is ~0.003 cm, that of a metastable particle is ~144 cm.

Plasma sterilization

There are three traditional states of matter: liquid, gas and solid. Plasma may be considered the fourth state, in which the molecules of a gas are excited to become plasma when the gas atoms lose their electrons and generate a highly excited mixture of charged nuclei and free electrons. A true plasma is actually considered to consist of positively and negatively charged particles in approximately equal concentrations. Plasma can be generated by the application of sufficient energy, in the form of heat or an electromagnetic field, to a gas. A plasma can be subsequently formed by further energy absorption by the gas, which fragments the gas atoms and molecules to produce negative ions, positive ions, electrons, and other short-lived or long-lived reactive species (Fig. 2.3).

It should be remembered that an atom of any element consists of a central nucleus (made up of positively charged protons and neutrons) that is surrounded by negatively charged and paired electrons, which are organized in defined orbitals, depending on their energy levels. In this state, each atom is balanced, with an overall neutral charge produced by an equal number of electrons and protons. As energy is applied to the atoms/molecules in a gas, the molecules and atoms fragment to produce positive ions (as they now have a higher number of protons) and free, negatively charged electrons. In some cases, the electrons react with other atoms, thereby gaining an overall negative charge (negative ions). Further unstable species are also generated including ozone (in the case of oxygen plasmas; ozone causes a

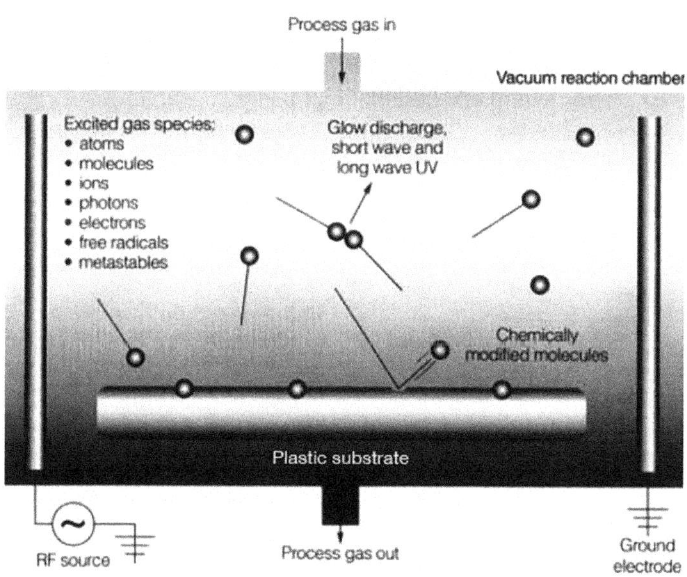

Figure 2.3 Model of the gas plasma exposure. Reproduced from URL of http://www.astp.com/plasma/pl_examples.html with permission from AST Products Inc.

serious etching phenomenon) and other free radicals, metastables, photons, UV and VUV. The free radicals that are formed include the hydroxyl radicals (OH·), NO radicals (NO·), oxygen metastables (O_2^1), or nitrogen metastables. Free radicals are highly reactive in that they have unpaired electrons in their outermost orbitals and therefore bind with electrons from other molecules to produce a chain reaction of electron loss and gain. Therefore, on exposure to microorganisms, a variety of effects occur, which cause functional damage to cell components (including proteins, lipids, nucleic acids and dipicolinic acid), ultimately resulting in cell death. The role of metastables in spore sterilization will be discussed in Chapter 4. Furthermore, with the excitation of electrons between atom orbitals, as they return to their natural states, they give off energy ($E = h\nu$) in the form of heat or photons, for example, within the UV wavelength range (~100 to 350 nm). This also contributes to antimicrobial activity as a minor factor. According to experimental results, sterilization effects of UV or VUV on spores are rarely observed, so UV-C is not an efficient contributor to gas plasma sterilization (Deng *et al.*, 2006). When the energy source applied to the gas is turned off, the various species rapidly recombine into lower-energy, stable forms.

A variety of plasmas can be produced, which are usually named after the gas used to create them, e.g. oxygen or nitrogen gas plasma. Several gases have been used for plasma generation including oxygen, nitrogen, argon, helium, and a mixture of oxygen and nitrogen at 1/4 (v/v). Most popular is the mixture of oxygen and nitrogen at 1/4 (v/v). Oxygen alone causes significant etching and shrinkage of microorganisms probably due to ozone, as mentioned above (Kylian *et al.*, 2006). Nitrogen and argon alone do not have significant effects on microorganisms, but the major contributors of nitrogen gas plasma are thought to be metastables of N_2 and O_2, so the presence of some O_2 is favourable; thus N_2/O_2 (4/1) is the most popular gas for sterilization. As discussed above, plasmas are generated by the application of heat or electromagnetic radiation. Heat is generally not used because of the very high temperatures and pressures required for the generation of plasmas (e.g. up to 3000°C). Lower-temperature plasmas are usually produced in a gas under vacuum with the application of microwaves or high-energy radiofrequency; in some cases atmospheric pressure plasma is also used (Shintani *et al.*, 2007, 2010), mostly for the purpose of surface modification (Fig. 2.2) as well as gas plasma sterilization (Fig. 2.3). These plasmas are usually generated under relatively low pressure such as half to one-third pressure at low temperatures (30 to 50°C) (Shintani *et al.*, 2007, 2010).

Application

Plasmas can be used in sterilization processes (Shintani *et al.*, 2007, 2010) with potential applications in liquid waste disposal, water disinfection, and surface and air disinfection. While presently being considered for use for medical device and medical material sterilization in healthcare and industrial applications, plasma sterilization is currently not used outside of research settings. The methods used to generate the plasma vary: in some cases, plasma is generated using the gas within a given chamber, while in others the plasma can be created in a separate chamber and introduced into the sterilization chamber. The latter method is being used by the NGK Co. Ltd (Nagoya, Japan) to develop its Remote Plasma sterilizer. Plasmas have also been described for the generation of ozone (limited to O_2 gas plasma) and other reactive species from oxygen. Oxygen gas plasma causes significant etching and shrinkage (Lee *et al.*, 2006; Rossi and Kylian, 2012), resulting in failure of material/

Table 2.1 Dissociation energy of several gases

Gas	F_2	H_2O_2	OH	N_2O	O_2	CO_2	NO	N_2
Dissociation energy (eV)	1.66	2.21	4.62	4.93	5.21	5.52	6.50	9.91

functional compatibility. Therefore, oxygen gas plasma is not recommended for gas plasma sterilization.

Recent applications have involved plasma production at room temperature and at atmospheric (or slightly negative) pressure by dielectric barrier discharge. This is achieved by the passage of a gas through a pair of electrodes, which are covered by a dielectric material (or simply an insulating or non-conductive material like quartz glass plates), which prevents arcing when a current is applied. In addition to gases, energy can be applied to liquids, including water, which can also cause similar dissociation of atoms or molecules (Table 2.1). Although these may not be considered true plasmas, they are similar to the generation of electrolysed or activated water, which can destroy waterborne microorganisms (Hayes *et al.*, 2013; Rowan *et al.*, 2008) or VBNC (viable but non-culturable) microorganisms (Brelles-Marino, 2012; Cooper *et al.*, 2010).

In general, plasmas demonstrate broad-spectrum antimicrobial activity due to the production of many reactive species; not surprisingly, bacterial spores (particularly spores from aerobic bacteria, including those from *Bacillus* spp. and *Geobacillus* spp. (Shintani *et al.*, 2007, 2010) demonstrate the greatest resistance (Table 2.2), with longer exposure times required for sterilization (Shintani *et al.*, 2007, 2010). In our case, the *D* value (decimal reduction value, time or dose to reduce one log) was approximately 1–1.2 min for *Geobacillus stearothermophilus* ATCC 7953 (Shintani *et al.*, 2007, 2010). The potency of the plasma depends on the vacuum applied and the gas used to generate the plasma. Most of them with some exception are, however, short-lived, which means they should be generated and applied close to the surfaces being treated; for these reasons, they are non-penetrating. Antimicrobial processes of plasmas are generally rapid due to their reactive nature and the fact that little or no residue remains on surfaces following treatment and simple aeration. Even though they are reactive, plasmas are not damaging to various metal and plastic surfaces because the penetration depth of the reactive species of the plasma is quite shallow (~10–20 nm) (Shintani, *et al.*, 2007).

Although the exact modes of action of plasmas are not yet well understood, it is known that the reactive species in a typical plasma react with various cell surface molecules and, potentially, internal proteins, nucleic acids (Brock, 2014; Brun *et al.*, 2012) and other essential molecules such as dipicolinic acid (Chapter 2). Ions and electrons are not considered major factors for sterilization because the outer surface of spores, as well as those of Gram-negative and Gram-positive microorganisms are charged (McDonnell, 2007), so ionic species are trapped at the outer layer and cannot penetrate into the interior. The most likely contributors to the sterilization process are metastables, rather than radicals (Popov, 2011; Takamatsu *et al.*, 2014; Guerra *et al.*, 2001; Ono *et al.*, 2009; Yagyu *et al.*, 2009; Bourig *et al.*, 2007; Vagin *et al.*, 2006; Yu *et al.*, 2006). This is because radicals are quite short-lived (μs) and have a short flight distance (0.003 cm/μs), in contrast to metastables, which have a relatively long life period (7 s to 2 s) and a longer flight distance (144 cm/2 s). UV and VUV are not major contributors to gas plasma sterilization based on current research (Deng *et al.*, 2006).

Table 2.2 Relative resistance to gas plasma exposure

	Microorganism	Examples
More resistant ↑	Prions	Scrapie, Creutzfeldt–Jakob disease, chronic wasting disease
	Bacterial spores	*Bacillus*, *Geobacillus*, *Clostridium*
	Prozoal oocysts	*Cryptospordium*
	Helminth eggs	*Ascaris*, *Enterobius*
	Mycobacteria	*Mycobacterium tuberculosis*, *M. terrae*, *M. chelonae*
	Small, non-enveloped viruses	Poliovirus, parvovirus, papillomaviruses
	Protozoal cysts	*Giardia*, *Acanthamoeba*
	Fungal spores	*Aspergillus*, *Penicillium*
	Gram-negative bacteria	*Pseudomonas*, *Providencia*, *Escherichia*
	Vegetative fungi and algae	*Aspergillus*, *Trichophyton*, *Candida*, *Chlamydomonas*
	Vegetative helminths and protozoa	*Ascaris*, *Cryptosporidium*, *Giardia*
	Large, non-enveloped viruses	Adenoviruses, rotaviruses
	Gram-positive bacteria	*Staphylococcus*, *Streptococcus*, *Enterococcus*
Less resistant	Enveloped viruses	Human immunodeficiency virus, hepatitis B virus, herpes simplex virus

Reproduced from McDonnell G.E. (2007) Antisepsis, Disinfection, and Sterilization, Types, Action, and Resistance, ASM Press, NW, Washington DC with permission from ASM Press.

Plasma sterilization can be used with a biological indicator (BI) such as *Geobacillus stearothermophilus* ATCC 7953 spores, because the bacterial endospore is the most tolerant to gas plasma (Table 2.2). Therefore, if spores of the BI are killed, other microorganisms in the bioburden are also expected to be killed. The order of tolerance to gas plasma exposure is presented in Table 2.2. Sterilization is defined as complete killing of spores and vegetative cells, whereas disinfection is defined as death of all vegetative cells but not spores. Therefore, sterilization is more stringent in terms of killing microorganisms (Shintani and McDonnell, 2011).

Sterilization and material compatibility

Plasma technology has also been considered for disinfection and sterilization of medical devices. The most favourable aspect of plasma technology is the possibility for simultaneous surface modification and sterilization in biomedical device fabrication (Figs. 2.2 and 2.3). Plasma sterilization may be suitable for medical implants and devices that are sensitive to temperature, radiation and chemicals. Plasma penetration depth is quite shallow at around ~10–20 nm; therefore material/functional compatibility can be easily attained together with a sterility assurance level (SAL) of 10^{-6} (Shintani *et al.*, 2007, 2010). This will be discussed in detail in Chapter 3.

Conclusion

Gas plasma sterilization was first tested in 1970, so this technology has a relatively short history. Since 2000, studies of gas plasma have expanded significantly. The most popular gas plasma equipment is low-temperature non-equilibrium equipment (Lassen *et al.*, 2006; Alfa *et al.*, 1996; Crow *et al.*, 1995) using a gas mixture of N_2/O_2 (4/1, v/v). This is because both a sterility assurance level of 10^{-6} and material/functional compatibility can easily be attained with this combination. To date, however, the mechanism of gas plasma sterilization remains unclear; after elucidation of the mechanism, there will be high potential for innovation of the types of appropriate gases and functional conditions to improve gas plasma sterilization technology.

References

Alfa, M.J., DeGagne, P., Olson, N., and Puchalski, T. (1996). Comparison of ion plasma, vaporized hydrogen peroxide, and 100% ethylene oxide sterilizers to the 12/88 ethylene oxide gas sterilizer. Infect. Control Hosp. Epidemiol. *17*, 92–100.

Ascham, L.E., and Menashi, W.P. (1972). Treatment of surface with low pressure plasmas. US Patent 3,701,628.

Bouring, A., Lago, V., Martin, J.-P., Pliavaka, K., Pliavaka, F., Gorbatov, S., and Chernukho, A. (2007). Generation of singlet oxygen in HV pulsed + DC crossed discharge at atmospheric pressure for oxygen-enhanced combustion. Int. J. Plasma Environment. Sci. Technol. *1*, 57–67.

Brelles-Marino, G. (2012). Induction of viable-but-non-culturable state in bacteria treated with gas discharge plasma. J. Appl. Microbiol. *112*, 412–413.

Brock, T. (2014). DNA damage: Roles in diseases and therapies, SFRBM 2014.

Brun, P., Vono, M., Venier, P., Tarricone, E., Deligianni, V., Martines, E., Zuin, M., Spagnolo, S., Cavazzana, R., Cardin, R., Castagliuolo, I., Valerio, A.L., and Leonardi, A. (2012). Disinfection of ocular cells and tissues by atmospheric pressure cold plasma. PLoS One 7, e33245.

Cooper, M., Friedman, G., Friedman, A., and Joshi, S.G. (2010). Biological responses of *Bacillus stratosphericus* to floating electrode–dielectric barrier discharge plasma treatment. J. Appl. Microbiol. *109*, 2039–2048.

Crow, S., and Smith, J.H. 3rd. (1995). Gas plasma sterilization-application of space and technology. Infect. Control Hosp. Epidemiol. *16*, 483–187.

Deng, X., Shi, J., and Kong, M.G. (2006). Physical mechanisms of inactivation of *Bacillus subtilis* spores using cold atmospheric plasmas. IEEE Trans. Plasma Sci. *34*, 1310–1316.

Guerra, V., Sa, P.A., and Loureiro, J. (2001). Role played by the N_2 metastable in stationary N_2 and N_2-O_2 discharges. J. Phys. D Appl. Phys. *34*, 1745–1755.

Hayes, J., Kirf, D., Garvey, M., and Rowan, N. (2013). Disinfection and toxicological assessments of pulsed UV and pulsed-plasma gas-discharge treated-water containing the waterborne protozoan enteroparasite *Cryptosporidium parvum*. J. Microbiol. Methods *94*, 325–337.

Kylian, O., Sasaki, T., and Rossi, F. (2006). Plasma sterilization of *Geobacillus stearothermophilus* by O_2;N_2 RF inductively coupled plasma. Eur. Phys. J. Appl. Physics. *34*, 139–142.

Lassen, K.S., Nordby, B., and Gruen, R. (2003). Optimization of a RF-generated CF4/O2 gas plasma sterilization process. J. Biomed. Mater. Res. *65*, 239–244.

Lassen, K.S., Johansen, J.E., and Gruen, R. (2006). Comparison of two radio-frequency plasma sterilization processes using microspot evaluation of microbial inactivation. J. Biomed. Mat. Res. *78*, 189–195.

Lee, K., Peak, K.-H., Ju, W.-T., and Lee, Y. (2006). Sterilization of bacteria, yeast, and bacterial endospore by atmospheric pressure cold plasma using helium and oxygen. J. Microbiol. *44*, 269–275.

McDonnell, G.E. (2007). Antisepsis, Disinfection, and Sterilization (American Society of Microbiology Press).

Ono, R., Tobaru, C., Teramoto, Y., and Oda, T. (2009). Plasma Source Sci. Technol. *18*, 025006.

Popov, N.A. (2011). Effect of singlet oxygen O_2 molecules produced in a gas discharge plasma on the ignition of hydrogen–oxygen mixtures. Plasma Sources Sci. Technol. *20*, 045002.

Rossi, F., and Kylian, O. (2012). Sterilization and decontamination of surfaces by plasma discharges. In Sterilization of biomaterials and medical devices. Lerouge, S and Simmons, A. eds. (Woodhead Publishing), Cambridge, UK, pp. 117–150.

Rowan, N.J., Espie, S., Harrower, J., Earrell, H., Marsill, L., Anderson, J.G., and MacGregor, S.J. (2008). Evidence of lethal and sublethal injury in food-borne bacterial pathogens exposed to high-intensity pilsed-plasma gas discharges. Lett. Appl. Microbiol. *46*, 80–86.

Shintani, H., and McDonnell, G.E. (2011). Definition of sterilization, disinfection, decontamination, and antisepsis. In Sterilization and Disinfection by Plasma, Sterilization Mechanisms, Biological Applications, Sakudo, A., and Shintani, H. eds (NOVA Science Publishers Inc.), New York, pp. 7–18.

Shintani, H., Shimizu, N., Imanishi, Y., Sekiya, T., Tamazawa, K., Taniguchi, A., and Kido, N. (2007). Inactivation of microorganisms and endotoxins by low temperature nitrogen gas plasma exposure. Biocontrol Sci. *12*, 131–143.

Shintani, H., Sakudo, A., Burke, P., and McDonnell, G.E. (2010). Gas plasma sterilization of microorganisms and mechanisms of action. Exp. Ther. Med. *1*, 731–738.

Takamatsu, T., Uehara, Kodai., Sasaki, Y., Miyahara, H., Matsumura, Y., Iwasawa, A., Ito, N., Azuma, T., Kohno, M., and Okino, A. (2014). Investigation of reactive species using various gas plasmas. RSC Adv. *4*, 39901–39905.

Vagin, A.A., Ionin, Yu.M., Klimachev, A.A., Kitkov, I.V., Kochetov, A.P., Napartovich, Yagyu, Y., Hayashi, N., Guan, W., Kawasaki, H., Ohshima, T., and Suda, Y. (2009). Influence of atomic and singlet molecular oxygen generated by RF plasma on reduction of protein. J. Plasma Fusion Res. *8*, 578–584.

Yu, O., Podmar'kov, O.A., Rulev, L.V., Seleznev, D.V., Sinitsyn, M.P., Frolov, N.N., and Yuryshev, N.N. (2006). Singlet oxygen in the low-temperature plasma of an electron-beam-sustained discharge. Plasma Phys. Rep. *32*, 429–429.

Concomitant Achievement of a Sterility Assurance Level of 10^{-6} with Material and Functional Compatibility by Gas Plasma Sterilization

3

Hideharu Shintani

Abstract

In this chapter, we discuss the importance of attaining a sterility assurance level (SAL) of 10^{-6} while maintaining material/functional compatibility. Simultaneous achievement of both these factors is required in ISO 14161 and for sterilization validation. Because the level of penetration achieved by gas plasma sterilization is quite shallow, at around 10–20 nm from the surface, the procedure kills only one layer of bioburden but readily maintains material and functional compatibility. In reality, there is an absence of bioburden that form multilayer clumps in healthcare products. Thus, gas plasma treatment easily displays material/functional compatibility whilst achieving a SAL of 10^{-6} due to its low temperature of operation and shallow penetration.

Introduction

Gas plasma sterilization is popular among sterilization researchers and a small number of commercial gas plasma sterilizers are available from, for example AST Products Inc. (http://www.astp.com/plasma-equipment). However, gas plasma sterilization is not popular due to the narrow space of the sterilization chamber. Sterilization represents the most rigorous process to eliminate microorganisms (see Table 2.2). Sterilization can kill all types of microorganisms including spores and vegetative cells (Sakudo and Shintani, 2011). Spores are the most tolerant of all microorganisms (see Table 2.2). In addition, according to ISO 14161 and sterilization validation, proper sterilization must attain a sterility assurance level (SAL) of 10^{-6} with an initial population of 10^6 CFU (colony-forming units). Reduction from an initial population down to a SAL of 10^{-6} requires a 12 log reduction. The requirement of a '6 log reduction' is not a reduction of an initial population of 10^6 CFU/carrier to 10^0 CFU/carrier. The correct 6 log reduction required by the authority is from 10^0 CFU/carrier (bioburden level) to a SAL of 10^{-6} as described in ISO 14161 and sterilization validation. For this purpose a straight survivor curve is required. An initial population of 10^0 CFU/carrier is a reasonable population expected for the bioburden and a SAL of 10^{-6} is specifically required in ISO 11138-1 and ISO 14161 as well as for sterilization validation. The 6 log reduction required for BI users is the absolute bioburden method in ISO 14161. This requirement is not addressed to BI manufacturers in ISO 11138-1. Further details of these methods will be discussed in Chapter 13.

Requirements for the sterilization procedure

The requirements for sterilization validation are provided in ISO 14161 and 11138-1. If a straight line survivor curve can be experimentally demonstrated for the reduction of an initial population (10^6 CFU/carrier) to a SAL of 10^{-2}, then a SAL of 10^{-2} to a SAL of 10^{-6} can be speculated to be a straight line (Fig. 3.1). This reduction cannot be confirmed experimentally and can only be speculated from the stochastics in ISO 11137-1. A SAL of up to 10^{-2} (1/100) can be confirmed experimentally, but a SAL less than 10^{-3} has more chance to have contamination. Therefore exact SALs of less than 10^{-2} remain uncertain and a SAL of 10^{-6} is actually speculation. This amount is defined as the closest amount to zero based on stochastics, a concept that is explained in ISO 11137-1. Any tailing in the reduction of an initial population of 10^6 CFU/carrier to a SAL of 10^0 due to clumping of the BI is inappropriate (see Fig. 1.1) (curved survivor line; see Fig. 1.3A), and means that the data are not valid. The reason why tailing curves are observed and how to avoid them is also explained in the NOVA book (Sakudo and Shintani, 2011). However, curved survivor lines even for SALs of 10^2–10^3 can be observed in the papers and books from engineering researchers. All of these data are invalid.

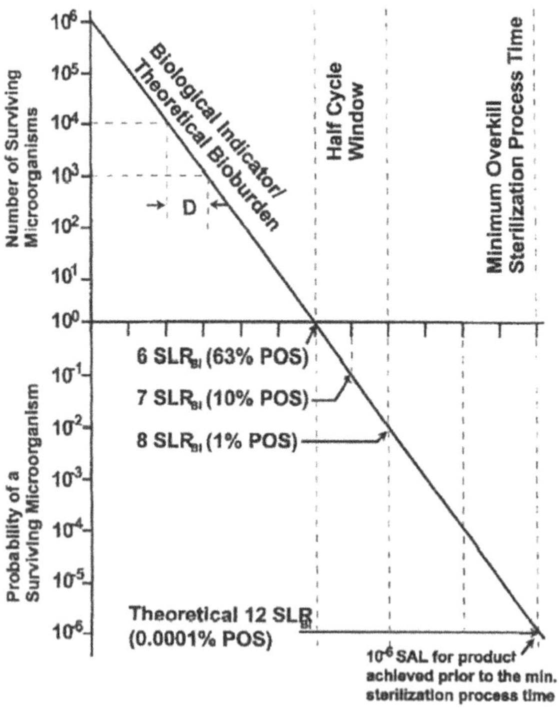

Figure 3.1 Necessity to attain a SAL of 10^{-6} and straight survivor curve from the initial population to a SAL of 10–6. SLR stands spore log reduction, POS stands possibility of survival. Reproduced from Shintani (2015).

Reduction of an initial population of 10^6 CFU/carrier down to a SAL of 10^{-6} is quite difficult to attain by gas plasma sterilization because the penetration depth of gas plasma is quite shallow, ~ 10 to 20 nm (Shintani *et al.*, 2007). Fig. 3.2 shows the polystyrene surface after (upper) and before (lower) gas plasma exposure as observed by atomic force microscopy (AFM). From the upper figure, the polystyrene can be observed to be etched to a depth of ~10 to 20 nm. From the presented scale the deepest etched depth can be estimated to be approximately 20 nm. *Geobacillus stearothermophilus* ATCC 7953 spores have a width of 1 μm and a length of 3 μm (Fig. 3.3), indicating that gas plasma cannot pass through even one spore. Therefore, gas plasma can kill only one layer of spores; multilayers are not efficiently killed because of the shallow penetration depth. The presence of multiple layers of the BI (clumping, see Fig. 1.1) is the reason why survival curves tail off before a SAL of 10^0 (Sakudo and Shintani, 2011). Multi-layers, known as clumping among microbiologists (see Fig. 1.1) and as stacking among engineering researchers, must be avoided to obtain a straight survival curve up to a SAL of 10^{-6}, not a SAL of 10^0 (Fig. 3.3).

As gas plasma sterilization has a very shallow penetration depth, products are generally quite safe from damage, indicating that simultaneous achievement of a SAL of 10^{-6} and

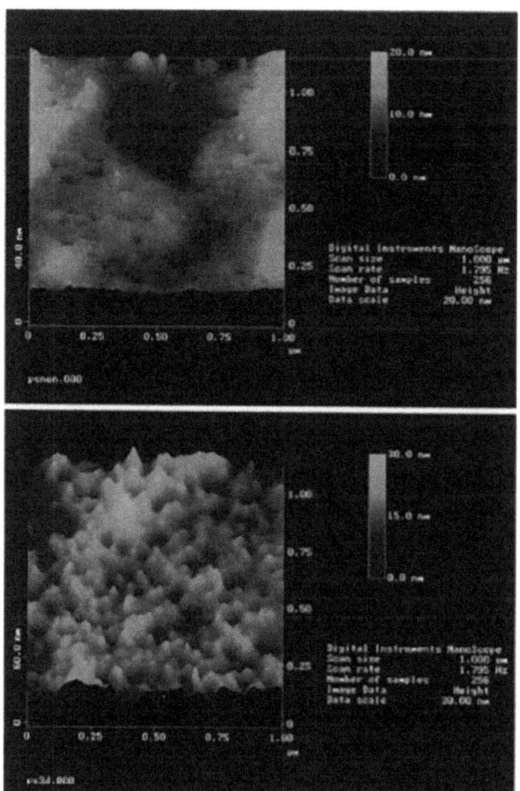

Figure 3.2 Surface analysis of atomic force microscopy (AFM) before and after gas plasma exposure to polystyrene (PS). Upper panel is after treatment for 7 min and lower panel is before treatment (control). Reproduced from Shintani *et al.* (2007).

Control

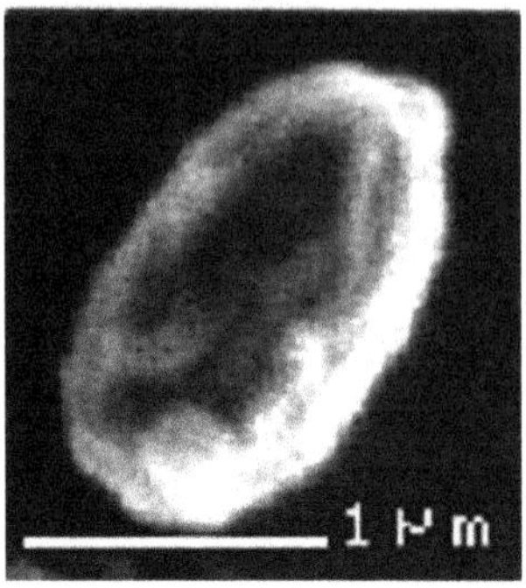

Figure 3.3 Scanning electron microscopy (SEM) of a spore of *Geobacillus stearothermophilus* ATCC 7953. Such spores are used as the biological indicator (BI) in gas plasma sterilization. Reproduced from Shintani *et al.* (2007).

material/functional compatibility can be easily obtained compared with existing sterilization procedures such as gamma-ray irradiation, autoclaving, dry heating, hydrogen peroxide gas sterilization, ethylene oxide gas sterilization, etc. (see Box 3.1).

We have data to indicate that gas plasma sterilization does not cause serious damage to many materials. Polystyrene (PS) was sterilized by nitrogen gas plasma and the amounts of CO, NOx, HCN, O_3 and N_2O before and after gas plasma exposure were determined. These amounts were individually determined and all were lower than existing safety levels, indicating that no significant deterioration of PS was observed (Table 3.1). FT-IR data before and

Box 3.1 List of publications citing that gas plasma sterilization simultaneously achieves an SAL of 10^{-6} and material/functional compatibility

Shintani, 1995; Bathina *et al.*, 1998; Feldman and Hui, 2014; Wod and Getty; Penna *et al.*, 1999; Du Pont; Rao, 2011; Kunishima, 2005; Willie *et al.*, 2004; Kim *et al.*, 2004; Brown *et al.*, 2002; Fisher *et al.*, 1997: Volny *et al.*, 2007; Kwok *et al.*, 2004; Williams *et al.*, 2004; Olde *et al.*, 2003; Lin *et al.*, 1995; Courtney *et al.*, 1978; Wood and Getty; Hauser *et al.*, 2011; Deilmann *et al.*, 2008; Hauser *et al.*, 2008; Simmons *et al.*, 2006; Grabow *et al.*, 2005; Charlebois *et al.*, 2003; Trostle *et al.*, 2002; McNulty *et al.*, 2002; Ferreira *et al.*, 2001; Liao *et al.*, 2001; Duffy *et al.*, 2000; Reeves *et al.*, 2000; Lewis and Nyman, 1999; Hury *et al.*, 1998; Collier *et al.*, 1996; Hesby *et al.*, 1997; Baier *et al.*, 1992: Haertel *et al.*, 2013; Aerts *et al.*, 2013; Isbary *et al.*, 2013; Lee *et al.*, 2013; Benetoli *et al.*, 2012; Ke *et al.*, 2011; Rederstoff *et al.*, 2011; Magureanu *et al.*, 2011; Yang *et al.*, 2010; Yuan *et al.*, 2010; Rainer *et al.*, 2010; Torres *et al.*, 2010: Guo *et al.*, 2008; Naseem *et al.*, 2004; Rederstorff *et al.*, 2011; Zhang, 2014; Wittenburg *et al.*, 2014; Delgado *et al.*, 2014; Vetten *et al.*, 2014; Pokorny *et al.*, 2012; Popoola *et al.*, 2010; Justan *et al.*, 2010; Kinnari *et al.*, 2010; Lleixa *et al.*, 2008; Peniston and Choi, 2007; Hopper *et al.*, 2003; McKellop *et al.*, 2000; Kuijpers *et al.*, 2000; Ibrahim *et al.*, 2012; Lee and Choi, 2012; Tessarolo, *et al.*, 2006; Muranyl *et al.*, 2010; Lerouge *et al.*, 2000; MacDonald *et al.*, 2012; Sheen *et al.*, 2008; Kvam *et al.*, 2012; Whittaker *et al.*, 2004; Yuen *et al.*, 2011: Gatineau *et al.*, 2012; Gao *et al.*, 2006; Zhang *et al.*, 2004; Yet-Pole *et al.*, 2004.

Table 3.1 Analysis of exhaust gas from polystyrene (PS) treated with nitrogen gas plasma

Procedure	Before and after treatment to PS	Detected gases (ppm)				
Low-pressure gas plasma	Sort of gases	CO[1]	NOx[2]	HCN[3]	O_3[4]	N_2O[5]
	Before	<2	<0.5	ND	ND	ND
	After	3.9	1.1	<0.1	<0.05	2.6

[1]UV-absorbance method.
[2]Chemical luminescence method.
[3]Pyrazolone light absorption method.
[4]Ozone detector.
[5]GC-MS.
ND, not detected, indicating less than limit of detection (LOD)
Reproduced from Shintani *et al.* (2015).

after gas plasma exposure to PS indicated that no significant change occurred (for example no oxidation at 1670/cm, which represents C=O functional groups) after treatment for 7 min (Fig. 3.4). In addition, we have results of nitrogen gas plasma exposure and autoclaving of scalpels (Figs. 3.5 and 3.6). These results indicate that the scalpel was unchanged before and after nitrogen gas plasma treatment for 8 min and 40 min (Fig. 3.5), but significant damage was observed after autoclaving for 15 min at 121.1°C (Fig. 3.6). Based on these results, we conclude that nitrogen gas plasma sterilization can successfully attain simultaneous achievement of a SAL of 10^{-6} and material and functional compatibility (Shintani *et al.*, 2007, Williams *et al.*, 2004; Volny *et al.*, 2007; Lin and Cooper, 1995; Kwok *et al.*, 2004; Kim *et al.*, 2004; Courtney *et al.*, 1978).

In addition, we carried out a tensile and elongation strength test of Latex rubber before and after nitrogen gas plasma exposure (Table 3.2) (Shintani *et al.*, 2007) and (Chin *et al.*, 2013; Strickler *et al.*, 2010; Lin *et al.*, 1991; Gentis *et al.*, 2013; DeHoff and Anusavics, 2009; Fisher and Stawarczyk, 2007; Kunze *et al.*, 2003; Niederer *et al.*, 1995; Collier *et al.*, 1996; Mishra *et al.*, 2003; Brown *et al.*, 2002) and leaching tests of latex rubber before and after nitrogen gas plasma exposure (Table 3.3) (Shintani *et al.*, 2007) and (Vollpracht and

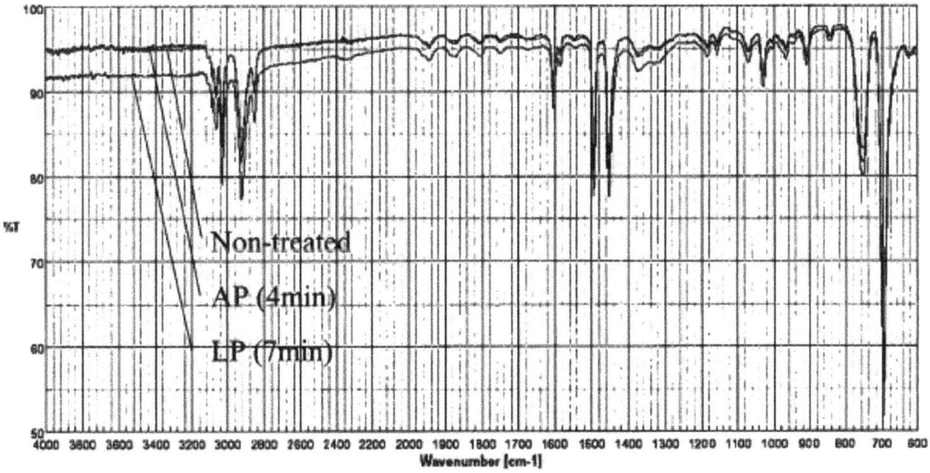

Figure 3.4 FT-IR data of PS before and after gas plasma exposure for 7 min treatment. AP, atmospheric pressure; LP, low pressure. Reproduced from Shintani *et al.* (2007).

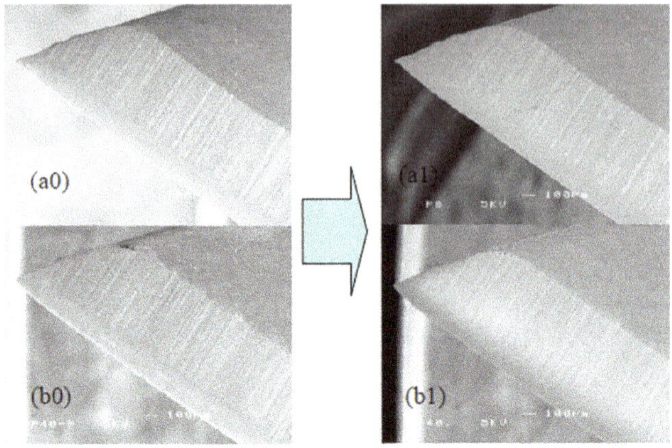

Figure 3.5 SEM observation of a scalpel before and after gas plasma exposure. (a0) is the control and (a1) is after 8 min treatment. (b1) is the control and (b1) is after 40 min exposure. Reproduced from Shintani *et al*. (2007).

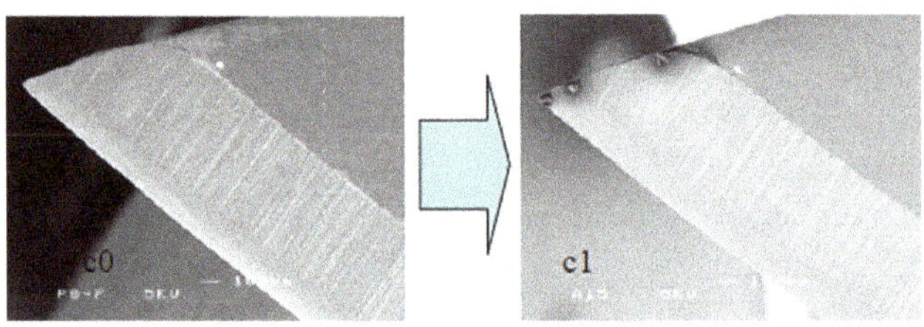

Figure 3.6 SEM observation of a scalpel before and after autoclave treatment for 15 min at 121.1°C. c1 is after treatment and c0 is the control.
Reproduced from Shintani *et al*. (2007).

Brameshuber, 2010; Bouvet *et al.*, 2007; van der Sloot *et al.*, 2001). Statistical analysis of data in Table 3.2 using the Student *t*-test (paired *t*-test using StatView^{R}) indicates that there was no significant difference. Statistical analysis of data in Table 3.3 cannot be done, but we can speculate that there is no significant difference.

Conclusion

As mentioned above, gas plasma sterilization can easily result in a SAL of 10^{-6} and material/functional compatibility. This is because the penetration depth of the sterilization factors is only 10–20 nm below the surface, which can kill scattered bioburden in a single layer on the products. Both achievement of a SAL of 10^{-6} and material/functional compatibility can be successfully achieved as required in ISO 14161 and sterilization validation by BI users. In many cases the existing sterilization procedures fail to attain material and functional compatibility; therefore, compatibility is not strictly applied to the existing sterilization procedures. If strictly applied, no compliant sterilization procedures would be currently available. Therefore, new sterilization procedures that meet all requirements to attain a SAL of 10^{-6} and material/functional compatibility must be developed.

Table 3.2 Tensile test, elongation test and 300% elongation test of Latex rubber before and after nitrogen gas plasma exposure

Sample	No.	Max. tensile test (N)	Max. elongation test (%)	300% Elongation tensile test (N)
No treatment	1	3.58	656.0	1.32
	2	3.83	684.5	1.09
	3	3.73	781.5	1.35
	4	3.73	695.0	1.33
	5	3.70	678.0	1.30
	Ave	3.71	699.0	1.28
Plasma, 40 min treatment	1	3.30	685.5	1.24
	2	3.45	694.5	1.25
	3	3.98	788.0	1.15
	4	3.80	626.5	1.43
	5	4.73	812.5	1.32
	Ave	3.85	721.4	1.08

Reproduced from Shintani *et al*. (2015).

Table 3.3 Leaching test of Latex rubber before and after of nitrogen gas plasma exposure

Sample	Heavy metal	Arsenic (As)	UV absorbance	Potassium permanganate	Evaporation residue
No treatment	Less than 2.0 ppm	Less than 2.0 ppm	220 nm: 0.3447 350 nm: 0.0562	8.90 µg/ml	2.0 mg
Plasma, 40 min exposure	Less than 2.0 ppm	Less than 2.0 ppm	220 nm: 0.3498 350 nm: 0.0479	9.22 µg/ml	3.7 mg

Reproduced from Shintani *et al*. (2015).

If gas plasma sterilization procedures were applied to healthcare materials in the future, the present exceptions to the existing sterilization procedures would no longer be approved by the authorities. There is an urgent need to improve existing sterilization procedures in order to attain a SAL of 10^{-6} and material and functional compatibility. Therefore, a commercial gas plasma sterilizer is needed as soon as possible.

Acknowledgements

This chapter is reproduced from a previous article (Hideharu Shintani, Pharmaceut Reg Affairs 4:131. doi: 10.4172/2167-7689.1000131) written by the same corresponding author (Hideharu Shintani) and distributed under the terms of the Creative Commons Attribution License (CCAL). It should be noted that CCAL permits unrestricted use, distribution and reproduction in any medium.

References

Aerts, R., Tu, X., Van Gaens, W., Whitehead, J.C., and Bogaerts, A. (2013). Gas purification by nonthermal plasma: a case study of ethylene. Environ. Sci. Technol. 47, 64758–66485.

Baier, R.E., Carter, J.M., Sorensen, S.E., Meyer, A.E., McGowan, B.D., and Kasprzak, S.A. (1992). Radiofrequency gas plasma (glow discharge) disinfection of dental operative instruments, including handpieces. J. Oral. Implantol. *18*, 236–242.

Bathina, M.N., Mikelsen, S., Brooks, C., Jaramillo, J., Hepton, T., and Kusumoto, F.M. (1998). Safety and efficacy of hydrogen peroxide plasma sterilization for repeated use of electrophysiology catheters. J. Am. Coll. Cardiol. *32*, 1384–1388.

Benetoli, L.O., Cadorin, B.M., Baldissarelli, V.Z., Geremias, R., de Souza, I.G., and Debacher, N.A. (2012). Pyrite-enhanced methylene blue degradation in non-thermal plasma water treatment reactor. J. Hazard Mater. 2012 Oct 30;237-238:55-62. Doi:10.1016

Bouvet, M., Francois, D., and Schwartz, C. (2007). Road soil retention of Pb leached from MSWI bottom ash. Waste Manag. *27*, 840–849.

Brown, S.A., Merritt, K., Woods, T.O., McNamee, S.G., and Hitchins, V.M. (2002). Effects of different disinfection and sterilization methods on tensile strength of materials used for single-use devices. Biomed. Instrument Technol. *36*, 23–27.

Charlebois, S.J., Daniels, A.U., and Lewis, G. (2003). Isothermal microcalorimetry; an analytical technique for assessing the dynamic chemical stability of UHMWPE. Biomaterials *24*, 291–296.

Chin, J.Y., Moon, K.W., Park, J.K., and Park, D.J. (2013). Development of reactive artificial liner recycled materials. 1. Mechanical properties and chemical compatibility. Waste. Manag. Res. *31*, 706–713.

Collier, J.P., Sutula, L.C., Currier, B.H., Currier, J.H., Wooding, R.E., Williams, I.R., Farber, K.B., and Mayor, M.B. (1996). Overview of polyethylene as a bearing material; comparison of sterilization methods. Clin. Orthop. Relat. Res. *333*, 76–86.

Courtney, J.M., Park, G.B., Prentice, C.R., Winchester, J.F., and Forbes, C.D. (1978). Polymer modification and blood compatibility. J. Bioeng. *2*, 241–249.

DeHoff, P.H., and Anusavice, K.J. (2009). Viscoelastic finite element stress analysis of the thermal compatibility of dental bilayer ceramic systems. Int. J. Prosthodont. *22*, 56–61.

Deilmann, M., Halfmann, H., Bibinov, N., Wunderlich, J., and Awakowicz, P. (2008). Low-pressure microwave plasma sterilization of polyethylene terephthalate bottles. J. Food Prot. *71*, 2119–2123.

Delgado, L.M., Pandit, A., and Zeugolis, D.I. (2014). Influence of sterilization methods on collagen-based devices stability and properties. Expert Rev. Med. Devices *11*, 305–314.

Duffy, R.E., Brown, S.E., Caldwell, K.L., Lubniewski, A., Anderson, N., Edelhauser, H., Holley, G., Tess, A., Divan, H., Helmy, M., Arduino, M., and Jarvis, W.R. (2000). An epidemic of corneal destruction caused by plasma gas sterilization. The Toxic Cell Destruction Syndrome Investigative Team. Arch. Ophthalnol. *118*, 1167–1176.

Du Pont. Tyvek[R] offers unmatched sterilization compatibility. http://www.dupont.com/products-and-services/packaging-materials-solutions/pharmaceutical-packaging/brands/tyvek-sterile-packaging/articles/start-sterile-stay-sterile.html

Feldman, L.A., and Hui, H.K. (2014). Compatibility of medical devices and materials with low-temperature hydrogen peroxide gas plasma. http://oldwww.epotek.com/SSCDocs/whitepapers/Tech%20Paper%2059.pdf

Ferreira, S.D., Dernell, W.S., Powers, B.E., Schochet, R.A., Kuntz, C.A., Withrow, S.J., and Wilkins, R.M. (2001). Effect of gas-plasma sterilization on the osteoinductive capacity of demineralized bone matrix. Clin. Orthop. Relat. Res. *388*, 233–239.

Fisher, J., and Stawarczyk, B. (2007). Compatibility of machined Ce-TZP/Al_2O_3 nanocomposite and a veneering ceramic. Dent. Mater. *23*, 1500–1505.

Fisher, J., Reeves, E.A., Isaac, G.H., Saum, K.A., and Sanford, W.M. (1997). Comparison of the wear of aged and non-aged ultrahigh molecular weight polyethylene sterilized by gamma irradiation and by gas plasma. J. Mater. Sci. Mater. Med. *8*, 375–378.

Gao, J., Yu, J., Li, Y., He, X., Bo, L., Pu, L., Yang, W., Lu, Q., and Yang, Z. (2006). Decoloration of aqueous Brilliant Green by using glow discharge electrolysis. J. Hazard Mater. *137*, 431–436.

Gatineau, M., El-Warrak, A.O., Bolliger, C., Mourez, M., and Berthiaume, F. (2012). Effects of sterilization with hydrogen @eroxide gas plasma, ethylene oxide, and steam on bioadhesive properties of nylon and polyethylene lines used for stabilization of canine stifle joint. Am. J. Vet. Res. 73, 1665–1669.

Gentis, N.D., Vranic, B.Z., and Betz, G. (2013). Assessing compressibility and compactibility of powder formulations with Near-Infrared Spectroscopy. Pharm. Dev. Technol. *18*, 156–171.

Grabow, N., Schlun, M., Sternberg, K., Hakansson, N., Kramer, S., and Hopper, Hopper, R.H. Jr., Young, A.M., Orishimo, K.F., and Engh, C.A. Jr. (2003). Effect of terminal sterilization with gas plasma or gamma radiation on wear of polyethylene liners. J. Bone Joint Surg. Am. *85*, 464–468.

Guo, Y., Liao, X., and Ye, D. (2008). Detection of hydroxy radical in plasma reaction on toluene removal [in Chinese]. J. Environ. Sci. *20*, 1429–1432.

Haertel, B., Strassenburg, S., Oehmigen, K., Wende, K., von Woedtke, T., and Lindequist, U. (2013). Differential influence of components resulting from atmospheric-pressure plasma on integrin expression of human HaCaT keratinocytes. Biomed. Res. Int. 2013;2013;761451. Doi:10.1155/2013/761451.

Hauser, J., Halfmann, H., Awakowicz, P., Koeller, M., and Essenwein, S.A. (2008). A double inductively coupled low-pressure plasma for sterilization of medical implant materials. Biomed. Tech. *63*, 199–203.

Hauser, J., Esenwein, S.A., Awakowicz, P., Steinau, H.U., Koeller, M., and Halfmann, H. (2011). Sterilization of heat-sensitive silicone implant material by low-pressure gas plasma. Biomed. Instrument. Technol. *45*, 75–79.

Hesby, R.M., Haganman, C.R., and Stanford, C.M. (1997). Effects of radiofrequency glow discharge on impression material surface wettability. J. Prosthet. Dent. *77*, 414–422.

Hury, S., Vidal, D.R., Desor, F., Pelletier, J., and Lagarde, T. (1998). A parametric study pf the destruction efficiency of Bacillus spores in low pressure oxygen-based plasmas. Lett. Appl. Microbiol. *26*, 417–421.

Ibrahim, N.A., Eld, B.M., Youssef, M.A., Ei-Sayed, S.A., and Salah, A.M. (2012). Functionalization of cellulose-containing fabrics by plasma and subsequent metal salt treatments. Carbohydr. Polym. *90*, 908–914.

Isbary, G., Shimizu, T., Li, Y.F., Stolz, W., Thomas, H.M., Morfill, G.E., and Zimmermann, J.L. (2013). Cold atmospheric plasma devices for medical issues. Expert Rev. Med. Devices, *10*, 367–377.

Justan, I., Tichy, F., and Slavicek, P. (2010). A new type of plasma knife and its effect on biological issue-a pilot study. Acta Chir. Plast. *52*, 31–34.

Ke, Z., Huang, Q., Zhang, H., and Yu, Z. (2011). Reduction and removal of aquous Cr (VI) by gas glow discharge plasma at the gas–solution interface. Environ. Sci. Technol. *45*, 7841–7847.

Kim, K., Kim, C., and Byun, Y. (2004). Biostability and biocompatibility of surface-grafted phospholipids monolayer on a solid substrate. Biomaterials *25*, 33–41.

Kinnari, T.J., Esreban, J., Zamora, N., Fernandez, R., Lopez-Santos, C.D., Yubero, F., Mariscal, D., Puertolas, J.A., and Gomez-Barrena, E. (2010). Effect of surface roughness and sterilization on bacterial adherence to ultra-high molecular weight polyethylene. Clin. Microbiol. Infect. *16*, 1036–1041.

Kuijpers, A.J., van Wachem, P.B., van Luyn, M.J., Plantinga, J.A., Engbers, G.H., Krijgsveld, J., Zaat, S.A., Dankert, J., and Feijen, J. (2000). In vivo compatibility and degradation of crosslinking gelatin gels incorporated in knitted Dacron. J. Biomed. Mater. Res. *51*, 136–145.

Kunishima, H. (2005). A comparative study of ethylene oxide, hydrogen peroxide gas plasma and low temperature steam formaldehyde sterilization. Inf. Control Hosp. Epidemiol. *26*, 486–489.

Kunze, C., Freier, T., Helwig, E., Sandner, B., Reif, D., Wutzier, A., and Radusch, H.J. (2003). Surface modification of tricalcium phosphate for improvement of the interfacial compatibility with biodegradable polymers. Biomaterials *24*, 967–974.

Kvam, E., Davis, B., Mondello, F., and Garner, A.L. (2012). Nonthermal atmospheric plasma rapidly disinfects multidrug-resistant microbes by inducing cell surface damage. Antimicrob. Agent Chemother. *56*, 2028–2036.

Kwok, S.C., Yang, P., Wang, J., Lie, X., and Chu, P.K. (2004). Hemocompatibility of nitrogen-doped, hydrogen-free diamond-like carbon prepared by nitrogen gas plasma immersion ion implantation-deposition. J. Biomed. Mater. Res. *70*, 107–114.

Lee, F.P., Wang, D.Y., Chen, L.K., Kung, C.M., Wu, Y.C., Ou, K.L., and Yu, C.H. (2013). Antibacterial nanostructured composite films for biomedical applications: microstructural characteristics, biocompatibility and antibacterial mechanisms. Biofouling *29*, 295–305.

Lee, J.K., and Choi, C.H. (2012). Two-stage reimplantation in infected total knee arthroplasrt using a re-sterilized tibial polyethylene insert and femoral component. J. Arthoplasty *27*, 1701–1706.

Lerouge, S., Wertheimer, M.R., Marchand, R., Tabrizian, M., and Yahia, L. (2000). Effect of gas composition on spore mortality and etching during low-pressure plasma sterilization. J. Biomed. Mater. Res. *51*, 128–135.

Lewis, G., and Nyman, J.S. (1999). A new method of determining the J-inteegral fracture toughness of very tough polymers; application to ultra high molecular weight polyethylene. J. Long. Term. Eff. Med. Implants. *9*, 289–301.

Liao, W.T., Lee, W.J., Chen, C.Y., and Shih, M. (2001). Decomposition of ethylene oxide in the RF plasma environment. Environment. Technol. *22*, 165–173.

Lin, J.C., and Cooper, S.L. (1995). Surface characterization and ex vivo blood compatibility study of plasma-modified small diameter tubing; effect of sulphur dioxide and hexamethyldisiloxane plasmas. Biomaterials *16*, 1017–1023.

Lin, S.Y., Lee, C.J., and Lin, Y.Y. (1991). The effect of plasticizers on compatibility, mechanical properties, and adhesion strength of drug-free Eudragit E. films. Pharm. Res. *8*, 1137–1143.

Lleixa Calvet, J., Grafahrend, D., Klee, D., and Moeller, M. (2008). Sterilization effect on starPEG coated polymer surfaces: characterization and cell viability. J. Mater. Sci. Mater. Med. *19*, 1631–12636.

MacDonald, D., Hanzlik, J., Sharkey, P., Parvizi, J., and Kurtz, S.M. (2012). In vivo oxidation and surface damage in retrieved ethylene oxide-sterilized total knee arthroplasties. Clin. Orthop. Relat. Res. *470*, 1826–1833.

McKellop, H., Shen, F.W., Lu, B., Campbell, P., and Salovey, R. (2000). Effect of sterilization method and other modifications on the wear resistance of acetabular cups made of ultra-high molecular weight polyetjhylene. A hip-simulator study. J. Bone Joint Surg. Am. *82A*, 1708–1725.

McNulty, D.E., Liao, Y.S., and Haas, B.D. (2002). The influence of sterilization method on wear performance of the low contact stress total knee system. Orthopedics *25*(2 Suppl.), s243–243.

Magureanu, M., Piroi, D., Mandache, N.B., David, V., Medvedovici, A., Bradu, C., and Parvulescu, V.I. (2011). Degradation of antibiotics in water by non-thermal plasma treatment. Water Res. *45*, 3407–3416.

Mishra, S., Viano, A., Fore, N., Lewis, G., and Ray, A. (2003). Influence of lamella features of UHMWPE on its physical and uniaxial tensile properties. I. Effect of sterilization method in uncrosslinked and unaged materials. Biomed. Mater. Eng. *13*, 135–146.

Muranyl, P., Wunderlich, J., and Langowski, H.C. (2010). Modification of bacterial structures by a low temperature gas plasma and influence on packing material. J. Appl. Microbiol. *109*, 1875–1885.

Naseem, A., Olliff, C.J., Martini, L.G., and Lioyd, A.W. (2004). Effects of plasma irradiation on the wettability and dissolution of compacts of griseofulvin. Int. J. Pharm. *269*, 443–450.

Niederer, P.F., Kaeser, R., Walz, F.H., Brunner, A., and Faerber, E. (1995). Compatibility considerations for low mass rigid-belt vehicles. Accid. Anal. Prev. *27*, 551–560.

Olde, R., Classe, M.B., Engbers, G.H., Grijpma, D.W., and Feijen, J. (2003). Gas plasma etching of PEO/`BT segmented block copolymers film. J. Biomed. Mater. Res. *65*, 417–428.

Peniston, S.J., and Choi, S.J. (2007). Effect of sterilization on the physicochemical properties of molded poly(L-lactic acid). J. Biomed. Mater. Res. B Appl Biomater. *80*, 67–77.

Penna, T.C., Ferraz, C.A., and Cassola, M.A. (1999). The presterilization microbial load on used medical devices and the effectiveness of hydrogen peroxide gas plasma against Bacillus subtilis spores. Infect. Control Hosp. Epidemiol. *20*, 465–472.

Pokorny, D., Slouf, M., and Fulin, P. (2012). Current knowledge on the effect of technology and sterilization on the structure, properties and longevity of UHMWPE in total joint replacement. Acta Chir. Orthop. Traumatol. Cech. *79*, 213–221.

Popoola, O.O., Yao, J.Q., Johson, T.S., and Blanchard, C.R. (2010). Wear, delamination, and fatigue resistance of melt-annealed highly crosslinked UHMWPE criciate-retaining knee inserts under activities of daily living. J. Orthop. Res. *28*, 1120–1126.

Rainer, A., Centola, M., Spadaccio, C., Gheradi, G., Genovese, J.A., Licoccis, S., and Trombetta, M. (2010). Comparative study of different techniques for the sterilization of poly-L-lactide electrospun microfibers; effectiveness vs. material degradation. Int. J. Artif. Organs 33, 76–85.

Rao, T.V. (2011). Gas Plasma Sterilization. www.slideshare.net/doctorrao/gasplasma-sterilization

Rederstorff, E., Fatimi, A., Sinquin, C., Ratiskol, J., Merceron, C., Vinatier, C., Weiss, P., and Collec-Jouault, S. (2011). Sterilization of exopolysaccarides produced by deeep-sea bacteria: impact on their stability and degradation. Mar. Drugs *9*, 224–241.

Reeves, E.A., Barton, D.C., FitzPatrick, D.P., and Fisher, J. (2000). Comparison of gas plasma and gamma irradiation in air sterilization on the delamination wear of the ultra-high molecular weight polyethylene used in knee replacements. Proc. Inst. Mech. Eng. H. *214*, 249–255.

Sakudo, A., and Shintani, H., eds. (2011). Sterilization and Disinfection by Plasma: Sterilization Mechanisms, Biological and Medical Applications (Nova Science Publishers, New York).

Schmitz, K.P. (2005). Mechanical properties of laser cut poly(L-lactide) micro-specimens; implantations for stent design, manufacture, and sterilization. J. Biomed. Eng. *127*, 25–31.

Shen, Y., Lei, L., Zhang, X., Zhou, M., and Zhang, Y. (2008). Effect of various gases and chemical catalysts on phenol degradation pathways by pulsed electrical discharges. J. Hazard Mater. *150*, 713–722.

Shintani, H. (1995). The relative safety of gamma-ray, autoclave, and ethylene oxide gas sterilization of thermosetting polyurethane, Biomed. Instrument Technol. *29*, 513–519.

Shintani, H. (2015). Simultaneous Achievement of Sterility Assurance Level (SAL) of 10^{-6} and Material and Functional Compatibility in Gas Plasma Sterilization. Pharmaceut. Reg. Affairs *4*, 131.

Shintani, H., Shimizu, N., Imanishi, V., Sekiya, T., Tamazawa, K., Taniguchi, A., and Kido, N. (2007). Inactivation of microorganisms and endotoxins by low temperature nitrogen gas plasma exposure, Biocontrol Sci. *12*, 131–143.

Shintani, H., Sakudo, A., Burke, P., and McDonnell, G. (2010). Gas plasma sterilization of microorganisms and mechanisms of action. Exp. Ther. Med. *1*, 731–738.

Simmons, A., Hyvarinen, J., and Poole-Warren, L. (2006). The effect of sterilization on a poly(dimethylsiloxine)/poly(hexamethylene oxide) mixed macrodiol-based polyurethane elastmer. Biomaterials, *27*, 4484–4487.

Strickler, F., Richard, R., McFadden, S., Lindquist, J., Schwarz, M.C., Faust, R., Wilson, G.J., and Boden, M. (2010). In vivo and in vitro characterization of poly(stylene-β-osobutylene-β-stylene) copolymer stent coating for biostability. Vascular compatibility and mechanical integrity. J. Biomed. Mater. Res. *92*, 773–782.

Tessarolo, F., Caola, I., Caciagli, P., Guarrera, G.M., and Nollo, G. (2006). Sterility and microbiological assessment of reused single-use cardiac electrophysiology catheters. Infect Control Hosp. Epidemiol. *27*, 1385–1392.

Torress, N., Oh, S., Appleford, M., Dean, D.D., Jorgensen, J.H., Ong, J.L., Agrawai, C.M., and Mani, G. (2010). Stability of antibacterial self-assembled monolayers on hydroxyapatite. Acta Biomater. *6*, 3242–3255.

Trostle, S.S., Hendrickson, D.A., and Franke, C. (2002). The effects of ethylene oxide and gas plasma sterilization on failure strength and failure mode of pre-tied monofilament ligature loops. Vet Surg. *31*, 281–284.

Van der Sloot, H.A., Hoede, D., Cresswell, D.J., and Barton, J.R. (2001). Leaching behavior of synthetic aggregates. Waste Manag. *21*, 221–228.

Vetten, M.A., Yah, C.S., Singh, T., and Gulumian, M. (2014). Challenges facing sterilization and depyrogenation of nanoparticles: Effects on structural stability and biomedical applications, Nanomedicine, *1000*, 1391–1399.

Vollpracht, A., and Brameshuber, W. (2010). Investigation on the leaching behavior of irrigated construction elements. Environ. Sci. Pollut. Res. Int. *17*, 1177–1182.

Volny, M., Elam, W.T., Ratner, B.D., and Turecek, K. (2007). Enhanced in-vitro blood compatibility of 316L stanless seel surfaces by reactive landing of hyaluronions. J. Biomed. Mater. Res. *80*, 505–510.

Whittaker, A.G., Graham, E.M., Baxter, R.L., Jones, A.C., Richardson, P.R., and Meek, G. (2004). Plasma cleaning of dental instruments. J. Hosp. Infect. *56*, 37–41.

Williams, R.L., Wilson, D.J., and Rhodes, N.P. (2004). Stability of plasma-treated silicone rubber and its influence on the interfacial aspects of blood compatibility. Biomaterials, *25*, 4659–4673.

Willie, B.M., Ashrafi, S., Alajbegovic, S., Burnett, T., and Bioebaum, R.D. (2004). Quantifying the effect of resin type and sterilization method on the degradation of ultrahigh molecular weight polyethylene after 4 years of real-time shelf aging. J. Biomed. Mater. Res. *69*, 477–489.

Wittenburg, G., Lauer, G., Oswald, S., Labudde, D., and Franz, C.M. (2014). Nanoscale topographic changes on sterilized glass surfaces affect cell adhesion and spreaing. J. Biomed. Mater. Res. *102*, 2755–2766.

Wood, L., and Getty J. Plasma processing for optimal medical device manufacturing. http://www.nordson.com/en-us/divisions/march/support/Literature/Documents/techpaper-medicalapplications1.pdf

Yang, X.L., Bal, M.D., and han, F. (2010). Rapid elimination of surface organophosphorous pollutants using hydroxy radical. Huan Jing Ke Xue. *31*, 1670–1674. Chinese.

Yet-Pole, I., Liu, Y.C., Han, K.Y., and She, T.C. (2004). Construction of a low-pressure microwave plasma reactor and its application in the treatment of volatile organic compounds. Environ. Sci. Technol. *38*, 3785–3791.

Yuan, M.H., Narengerile, Watanabe, T., and Chang, C.Y. (2010). DC water plasma at atmospheric pressure for the treatment of aqueous phenol. Environment. Sci. Technol. *44*, 4710–4715.

Yuen, P.K., Su, H., Gorai, V.N., and Fink, K.A. (2011). Three-dimentional interconnected microporous poly(dimethylsiloxane) microfluidic devices. Lab. Chip. *11*, 1541–1544.

Zhang, H., Yang, L., Yu, Z., and Huang, Q. (2014). Inactivation of *Microcystis aeruginosa* by DC glow discharge plasma; impacts on cell integrity, pigment contents and microcystins degradation. J. Hazard Mater. 2014 Mar 15:268:33–42. Doi: 10.1016

Zhang, R.B., Wu, Y., Li, G.F., Wang, N.H., and Li, J. (2004). Plasma induced degradation of indigo Carmine by bipolar pulsed dielectric barrier discharge (DBD) in the water–air mixture. J. Environ. Sci. (China) *16*, 808–812.

Current Progress in Advanced Technology of Nitrogen Gas Plasma for Remote Sterilization and Clarification of Sterilization

4

Hideharu Shintani, Naohiro Shimizu, Yuichiro Imanishi, Akikazu Sakudo, Takuya Uyama and Eiki Hotta

Abstract

In this chapter, we will introduce the advanced technology for remote sterilization by exposure to nitrogen gas plasma, which is generated by a pulsed power source. Sterilization was best achieved using nitrogen gas plasma at a relative humidity (RH) of 0.5%. Furthermore, sterilization efficiency was directly correlated with the levels of $OONO\cdot^-$ (peroxynitrite anion radical). These results suggest that $OONO\cdot^-$ is a major factor in the case of remote sterilization by nitrogen gas plasma.

Introduction

Several papers have been published on gas plasma sterilization (Sakudo and Shintani, 2012; Shintani *et al.*, 2007, 2010). The definition of sterilization can be found in Sakudo and Shintani (2012). The advantages of gas plasma are that sterilization with a sterility assurance level (SAL) of 10^{-6} and material/functional compatibility can be attained without any difficulty. This is because the penetration depth of gas plasma sterilization is quite shallow (10–20 nm; Shintani *et al.*, 2007) and therefore only one layer of bioburden can be sterilized. The bioburden represents the type and number of viable microorganisms existing in/on products. Most of the bioburden exists as one layer; therefore, deeper penetration capability is unnecessary for efficient sterilization.

In contrast, existing sterilization procedures including gamma-ray irradiation sterilization, electron-beam sterilization, moist heat sterilization, dry heat sterilization, ethylene oxide gas sterilization, hydrogen peroxide gas sterilization and so on, have the ability to penetrate deeper. Therefore, materials are easily sterilized using these methods and a SAL of 10^{-6} can be attained; however, the sterilized products are useless due to degradation of the product material during the sterilization process, a phenomenon called failure to attain material/functional compatibility (Shintani, 1995, 2014). Good manufacturing practice (GMP) and sterilization guidelines require simultaneous attainment of both a SAL of 10^{-6} and material/functional compatibility, but this requirement is difficult to attain with the existing sterilization procedures. Therefore, sterilization procedures using stable gases that are safe to handle and that have shallow penetration depths are needed. As described previously, gas plasma sterilization has the characteristics necessary to meet these requirements (Sakudo and Shintani, 2012; Shintani *et al.*, 2007, 2010).

Table 4.1 Brief description of experimental conditions

Types of gases	N_2, N_2+H_2O
Flow rate	2 l/min
Pressure	Atmosphere (1 atm)
Distance between reactor and treatment area	100 mm constant
Temperature of hotplate	55–75°C
Humidity	0.0–5.0% RH
Electric discharge of voltage	13–15 kV
Electric discharge of current	7–12 A
Repeated frequency	1 kHz

The relative humidity (RH) was measured at the upper and lower sites of the reactor with BKPRECISION Ltd., 725 digital temperature/humidity sensors. Exhaust NOx gases were determined using a Shimadzu NOA-7000 analyser and exhaust ozone gas was determined by using an EG-700EIII ozone monitor from Ebara Ltd.

For the studies introduced here, we used a remote type of nitrogen gas plasma sterilization procedure using a pulsed power supply of static induction thyristor (SIThy) (Shimizu, 2010; Uyama, 2015; Shintani *et al.*, 2015). Several factors associated with the sterilization procedure were determined, and the main factors associated with sterilization were identified (Table 4.1). In addition, appropriate sterilization conditions were identified and are reported herein.

Efficiency of atmospheric pressure nitrogen gas remote plasma sterilization

Samples to be sterilized, including a biological indicator (BI), were placed on a hotplate (Figs. 4.1–4.4). The distance between the reactor and hotplate was kept constant at 100 mm (Fig. 4.1). The temperature of the hotplate was varied from 55 to 75°C (Fig. 4.5).

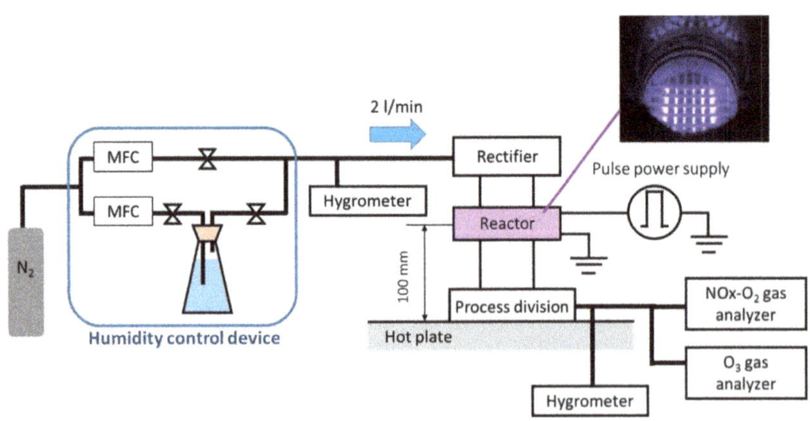

Figure 4.1 Schema of experimental system. The schema of the experimental system is shown. Remote gas plasma was utilized. The experimental system consists mainly of a humidity control device, plasma producer, and exhaust gas analyser. MFC stands for mass flow controller.

Figure 4.2 Plasma generation part. A photograph of the plasma generator parts is presented and the experimental conditions are presented in Table 4.1. A detailed explanation of the experimental conditions can be found in the Master's thesis of Takuya Uyama (TISTech, 2015).

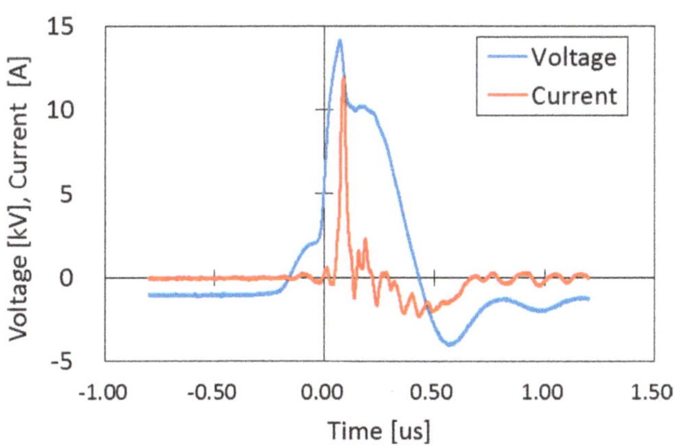

Figure 4.3 Waveform of voltage and that of current when plasma generated under the experimental conditions. The typical waveforms obtained when using the experimental conditions in Table 4.1.

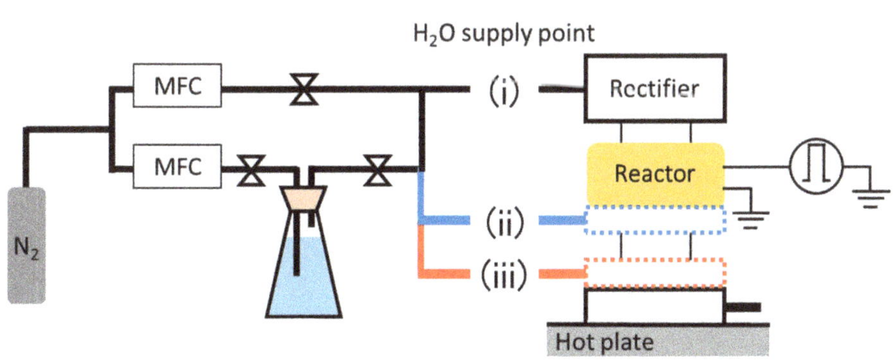

Figure 4.4 Water vapour supply locations

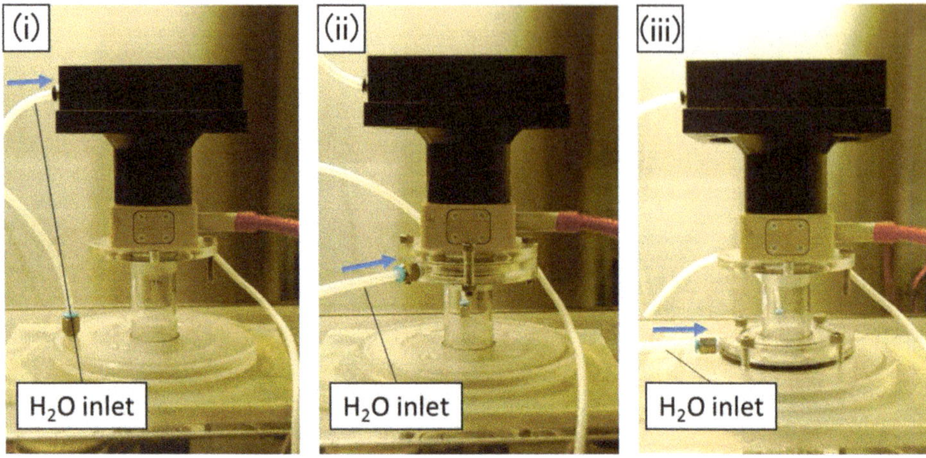

Figure 4.5 Relationship between sterilization efficiency and hotplate temperature. Relationship between sterilization efficiency and hotplate temperature is shown. It was determined that the higher the temperature, the greater the sterilization efficiency. It is seen that sterilization can be completed in 240 min, 150 min and 120 min at 55°C, 65°C and 75°C, respectively. These results indicate that 75°C is the best temperature for sterilization because an increase of 20°C from 55°C to 75°C resulted in a sterilization time that was half as long, and it is expected that the target materials including the BI are tolerant to this temperature. 75°C was found to be the optimal temperature.

Nitrogen gas was chosen for use in these sterilization studies because of its higher dissociation energy, which makes it relatively stable compared with other gases, and therefore it is inert and safe to handle (Table 4.2).

The need for humidity in gas plasma sterilization has been reported (Friedman and Friedman, 2013; Tamazawa *et al.*, 2015). A supply of water vapour was introduced at three locations as shown in Figs. 4.4 and 4.6. The site of water vapour introduction was varied because water vapour can play a role in generating various reactive oxygen species that may function in the sterilization process. Location (i) was at the upper part of the reactor, location (ii) was just below the reactor, and location (iii) was just before the site of sample treatment.

Table 4.2 Dissociation energy of several types of gases

Gas	Dissociation energy (eV)
N_2	9.91
O_2	5.21
H_2O	5.11
NO	6.50
SO_2	5.60
N_2O	4.93
CO_2	5.52
O_3	1.05
H_2O_2	2.21

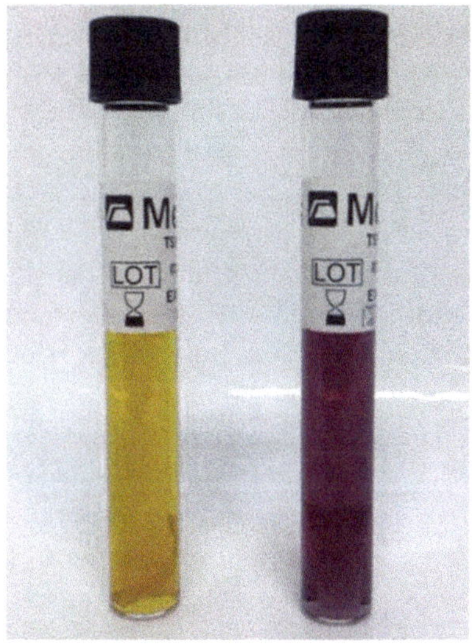

Figure 4.6 Photograph showing the differing location of water vapour supply.

At location (i), NO radicals, which are actively involved in the sterilization process, may be produced (Shintani *et al.*, 2014; Shintani, 2015). In location (ii), it can be speculated that N metastables or other reactive oxygen species (ROS) may be generated by reactions with water vapour. In location (iii), short-lived OH radical may attack the biological indicator (BI) and result in its sterilization. The actual experimental set up with the different positions of water vapour introduction is shown in Fig. 4.6.

Sterilization evaluation was confirmed by using a BI of *Geobacillus stearothermophilus* ATCC 7953 with 10^6 CFU (colony-forming unit)/carrier, which was obtained from MESA Lab. The *D* value (decimal reduction value) was obtained by two methods, the fraction negative method and survivor curve method (ISO 14161; ISO 11138-1). In the case of the fraction negative method, the BI was incubated using SCDB (soybean casein digest broth) liquid medium at 58°C for 2 days. The result was confirmed using a chemical indicator (CI; Fig. 4.7). When the BI survived, the colour changed to yellow, whereas when sterilization

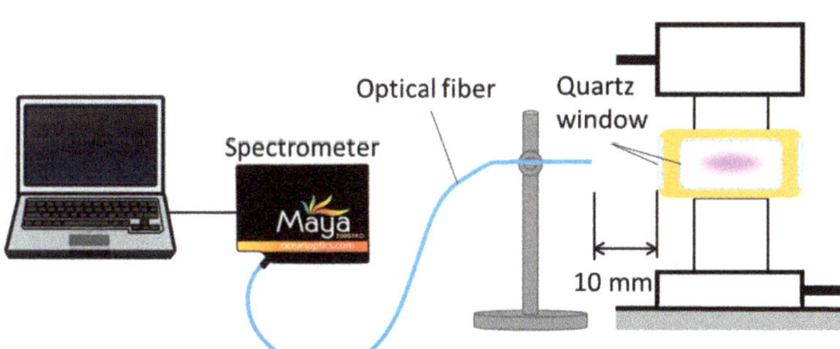

Figure 4.7 Colour change of chemical indicator (CI). The tube on the left indicates survival of the biological indicator (BI) (acid produced) and that on the right is sterilized (colour is unchanged).

was successful, the colour remained unchanged (purple). This is due to the production of organic acids (mostly citric acid) from the tricarboxylic acid (TCA) cycle (Fig. 4.7). To generate the survivor curve, we used SCDA (soybean casein digest agar) solid medium. Ten-fold serial dilutions using SCDB were carried to achieve final plate counts of 30–300 CFU/plate as required in ISO 14161. Spores were retrieved from the BI carrier by using the procedures described in ISO 11737-1. According to ISO 11138-1, the *D* value must be obtained using both the fraction negative method and the survivor curve method, so we carried out both methods following the ISO 11138-1 requirement.

The *D* value was determined by the Stumbo–Murphy–Cochran procedure, one of the fraction negative methods (ISO 14161). The results are summarized in Table 4.3. The *D* value was the lowest at a relative humidity (RH) of 0.5% (~ 8.7 min). The others were approximately 10 min, indicating that a RH of 0.5% resulted in the most efficient sterilization. *D* value was determined using the survival curve method. The *D* value was determined under the following conditions: hotplate temperature, 75°C, RH, 0.5% or 0%, and water vapour supply location (i) or (iii) as shown in Figs. 4.8 and 4.9.

From data in Fig. 4.8, it can be concluded that the use of a RH of 0.5% was superior to 0% RH, as the *D* was approximately 10 min. Results presented in Fig. 4.9 indicate that water vapour supply location (i) was superior to location (iii), and the *D* value was approximately 10 min. These data are consistent with the *D* values obtained by the fraction negative

Table 4.3 Determination of *D* value (min) by using fraction negative method, Stumbo-Murphy-Cochran Procedure and sterilization efficiency (%)

RH (%)	Non viable sheets	Viable sheets	Total sheets	*D* value (min)	Sterilization efficiency (%)
0.0	1	11	12	10.71	8.3
0.5	8	1	9	8.66	88.9
5.0	3	6	9	10.07	33.3

The treatment period was 60 min. Initial population was 10^6 CFU/carrier.

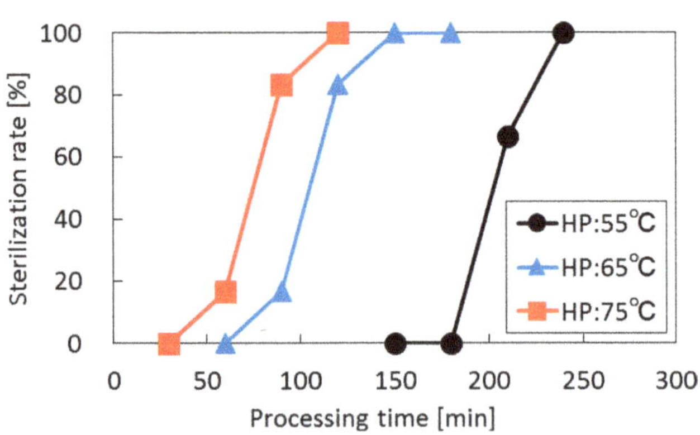

Figure 4.8 Relationship between treatment time and colony-forming units at water vapour supply portion of (i), RH of 0.5% and 0% and temperature at 75°C. 0.5% RH was found to be the most appropriate.

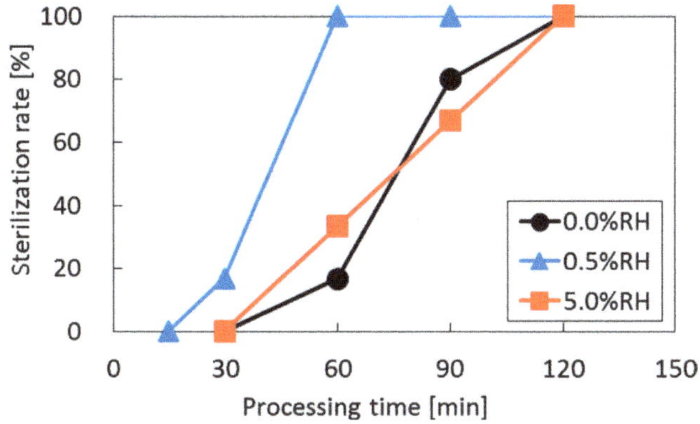

Figure 4.9 Relationship between exposure time and colony-forming unit at water vapour supply locations (i) and (iii), with a RH of 0.5% and a temperature of 75°C. Location (i) was found to be the most appropriate.

method. The *D* value from the survivor curve method was determined using a regression line with a coefficient of correlation of greater than 0.8 as required in ISO 11138-1.

Surfaces of spores were observed by using scanning electron microscopy (SEM; S-5500 Hitachi technologies Ltd). Fig. 4.10 shows the SEM observation of spores. Compared with the untreated control (left), sterilized spores showed no shrinkage, but some roughness of the surface was observed for spores that were treated for 30 min (middle). However, roughness did not always increase with increasing treatment time up to 90 min (right), indicating that roughness is a temporary rather than permanent phenomenon. It therefore appears that SEM observation does not provide any useful information regarding the success of the sterilization process. Nitrogen gas plasma does not cause any etching in contrast to O_2 gas plasma (Shintani *et al.*, 2010; Tamazawa *et al.*, 2015).

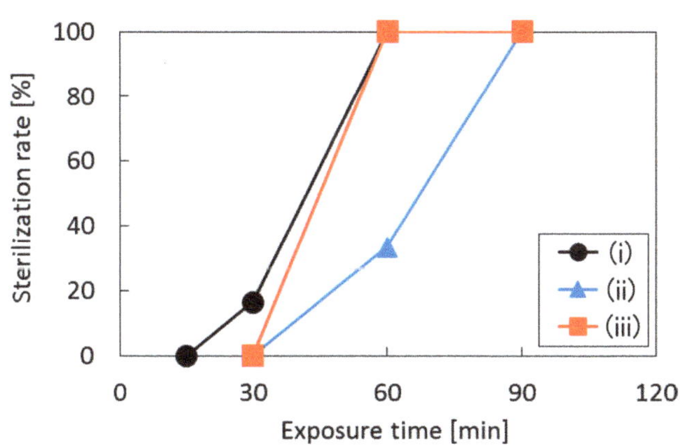

Figure 4.10 SEM observation of control and treated spores. The upper panel is the control, the middle panel is after a 30 min treatment and the lower panel is after a 90 min treatment. Roughness did not always increase with increasing treated time, indicating roughness is not always a permanent factor.

Clarification of sterilization major factors in the remote nitrogen gas plasma sterilization system

Several reactive oxygen species (ROSs) were analysed by using emission spectrophotometric analysers from Maya 2000 Pro (Ocean Optics Ltd) (Fig. 4.11). A quartz window was incorporated into the reactor and analyses were conducted under the following conditions. The determination wavelength was 200–650 nm, grating was 600 lines/mm, entrance slit width was 10 μm, exposure time was 100 ms and analyses were repeated five times.

In Fig. 4.12, the emission spectrum at a RH of 0.5% is shown. By using the equipment shown in Fig. 4.11, the emission spectrum can be obtained. NO radicals, N_2 second positives and N_2^+ were detected. However, no OH radicals were detected at 310 nm, indicating that OH radicals are not major contributors to nitrogen gas plasma sterilization.

A wavelength of 258.55 nm was used for the detection of NO radicals as shown in Fig. 4.12. In Fig. 4.13, the relationship between emission intensity at 258.55 nm and relative

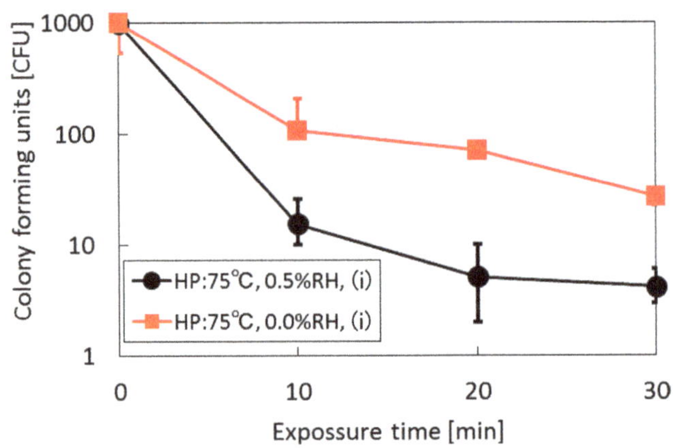

Figure 4.11 Analysis of reactive oxygen species (ROSs) by using emission spectrophotometric analysers.

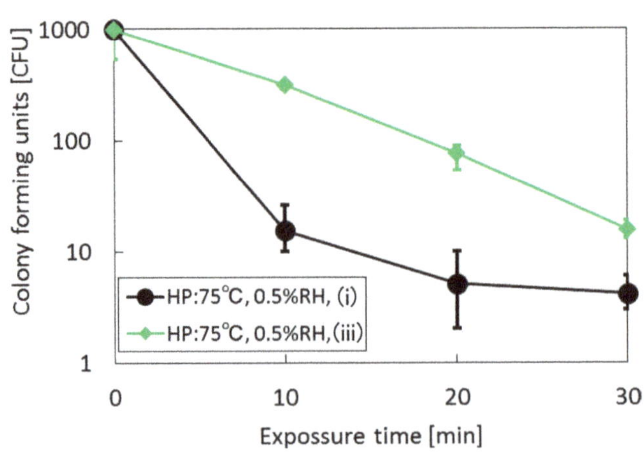

Figure 4.12 Emission spectrum at a RH of 0.5%. OH radical was not detected at 310 nm.

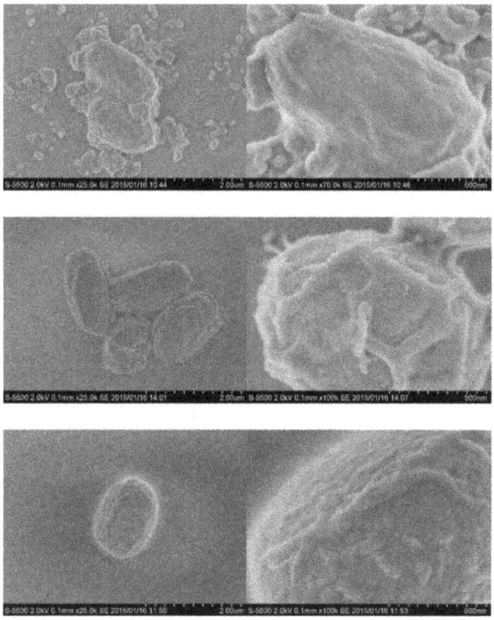

Figure 4.13 Relationship between emission intensity at 258.55 nm (NO radical formation) and relative humidity (RH).

humidity is presented. As shown in Table 4.3, sterilization efficiency was optimal at a RH of 0.5%, indicating that the NO radical itself does not function as a sterilization factor because results in Fig. 4.13 do not indicate that 0.5% RH was optimal. In addition, 258.55 nm is in the UV-C range and UV-C is thought to be effective for sterilization of microorganisms by causing thymine dimer formation. However, no role for UV-C in sterilization could be demonstrated in previous studies using *E. coli* and *Bacillus atrophaeus* ATCC 9372 (Li *et al.*, 2013; Deng *et al.*, 2006).

Production of one type of ROS, hydrogen peroxide (H_2O_2), has been reported by nitrogen gas plasma sterilization (Maeda *et al.*, 2015; Sakudo *et al.*, 2013, 2014), so we measured H_2O_2 by using a chemical indicator (CI). The CI for H_2O_2 analysis was from Quantofix Peroxide 25 (Macherey-Nagel Ltd) and the analysis range was 0–25 µg/ml. H_2O_2 formation was analysed using a CI from Macherey-Nagel Ltd, and the relationship between H_2O_2 concentration and humidity is presented in Fig. 4.14. As shown in Table 4.3, sterilization efficiency was optimal at a RH of 0.5%, indicating that H_2O_2 or OH radicals from H_2O_2 do not correlate with the RH tendency; therefore H_2O_2 or OH radicals do not appear to be major contributors in nitrogen gas plasma sterilization.

Superoxide anion radicals ($O_2\cdot^-$) were speculated to be produced at the reactor site and reach the treatment location as shown in Fig. 4.15. NO· and $O_2\cdot^-$ were also speculated to be produced even when the water vapour was introduced at location 3, the lowest part in Fig. 4.4, because sterilization was successful at this location.

Measurement of $O_2\cdot^-$ was not successful; therefore, its effect on gas plasma sterilization remains uncertain, but it can be speculated that $O_2\cdot^-$ supports the production of peroxynitrite anion radicals ($ONOO_2\cdot^-$), as shown in Fig. 4.16.

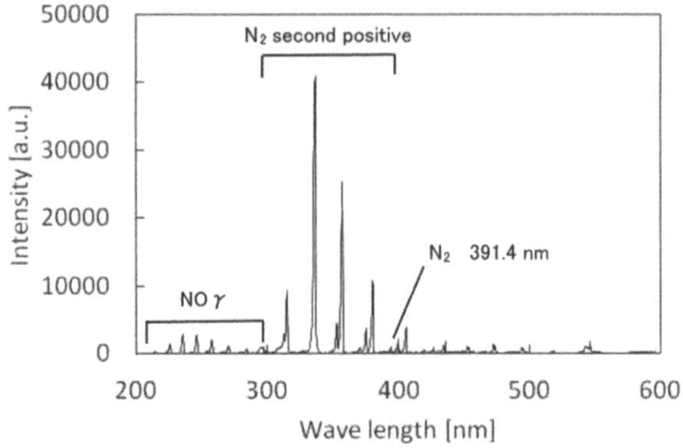

Figure 4.14 Relationship between H_2O_2 concentration and relative humidity (RH).

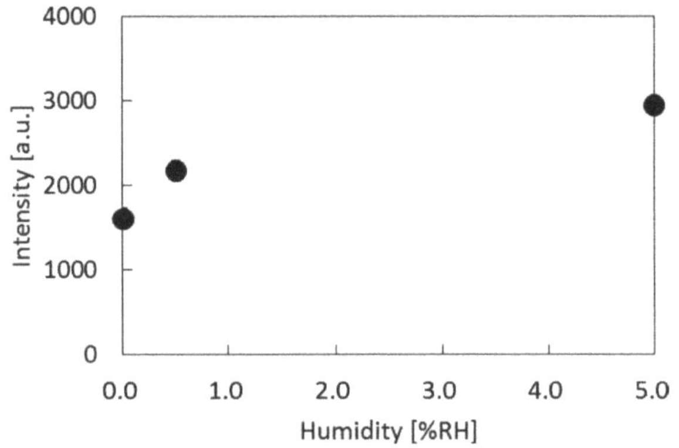

Figure 4.15 Speculated mechanism of formation of superoxide anion radical ($O_2{\cdot}^-$) in the reactor. $O_2{\cdot}^-$ produced in the reactor migrates to the treatment location and reacts with NO·, producing ONOO· $^-$ just before reaching the BI (Figs. 4.16 and 4.17).

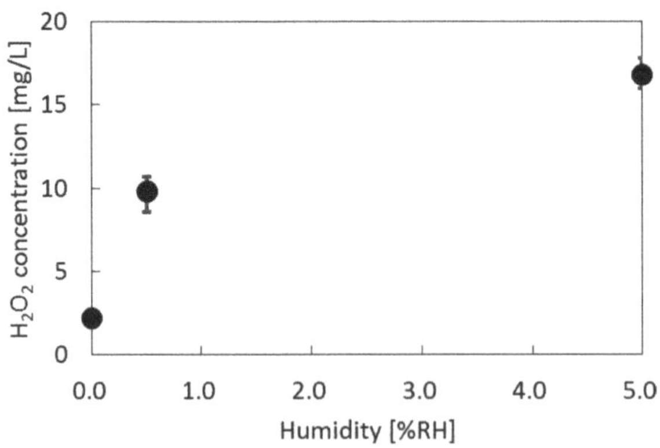

Figure 4.16 Production of peroxynitrite anion radical (ONOO·$^-$) from NO radical combines with superoxide anion radical ($O_2{\cdot}^-$). Peroxynitrite anion radical (ONOO·$^-$) causes nitration and hydroxylation, which cases deformation of protein or nucleic acids and results in sterilization. Nitration and hydroxylation are oxidation reactions. Peroxynitrite anion radical (ONOO$^-$) may be the precursor of OH radical if H^+ combines to ONOO·$^-$ and degrades to ·OH and $\cdot NO_2$.

OH radicals may be formed by the reaction shown in Fig. 4.16 and/or from H_2O_2. However, as shown in Fig. 4.12 and mentioned in section of 8, OH radicals were not detected and therefore OH radicals do not appear to be major contributors to nitrogen gas plasma sterilization.

Peroxynitrite anion radicals ($ONOO_2\cdot^-$) can be formed from NO radicals + superoxide anion radicals ($O_2\cdot^-$) (Fig. 4.16; Nova and Parola, 2008). The reaction in Fig. 4.16 will occur just at the upper layer of bacteria (Fig. 4.17), indicating that NO radicals and $O_2\cdot^-$ migrate from the reactor site to the treatment site and react as shown in Fig. 4.16 to produce $OONO_2\cdot^-$.

Peroxynitrite anion radicals were detected by using aminophenyl fluorescein (APF) reagent as shown in Fig. 4.18 (Setsukina *et al.*, 2003). The relationship between the peroxynitrite anion radical ($ONOO_2\cdot^-$) concentration and relative humidity is presented in Fig. 4.19. As shown in Table 4.3, sterilization efficiency was optimal at a RH of 0.5%, indicating that the peroxynitrite anion radical concentration correlates with the RH level. Based on this finding it can be speculated that peroxynitrite anion radicals function as a major sterilization factor in nitrogen gas plasma sterilization. In addition, please refer to the footnote of Fig. 4.18 for further clarification of $ONOO_2\cdot^-$ as a major sterilization factor.

Relationship between sterilization efficiency and water vapour supply location is shown in Fig. 4.20, water vapour supply locations (i) and (iii) in Fig. 4.4 were superior to that of (ii)

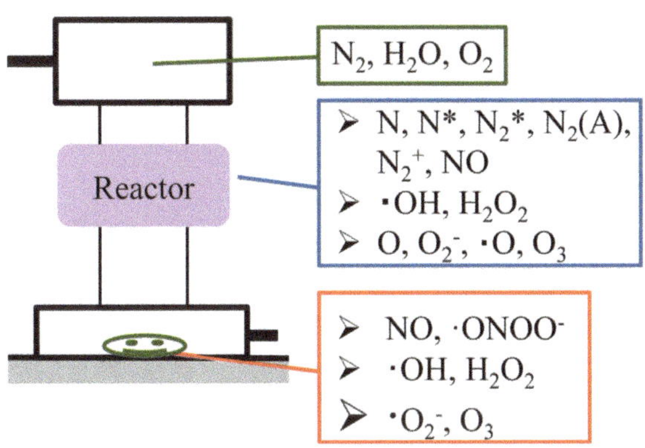

Figure 4.17 The series of reaction that produce HO·, $O_2\cdot^-$, NO· and ONOO·⁻ on the surface of bacteria.

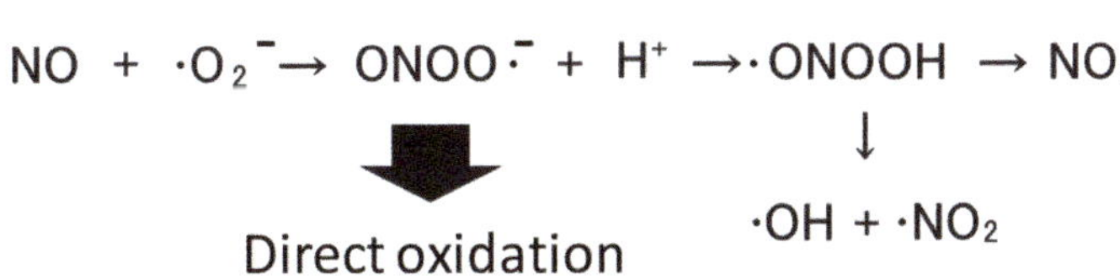

Figure 4.18 Peroxynitrite anion radical ($ONOO\cdot^-$) detection using aminophenyl fluorescein (APF). X is NH for APF and O for HPF (hydroxyphenyl fluorescein). hROS is a highly reactive oxygen species. APF is specific to OH radical and $ONOO\cdot^-$ and HPF is specific to OH radical. HPF does not detect OH radical, which indicates that the APF intensity is mostly due to $ONOO\cdot^-$ formation. When $ONOO\cdot^-$ combines with APF (the left reagent, X = NH), fluorescein compounds are produced at an excitation wavelength of 490 nm and emission wavelength of 515 nm.

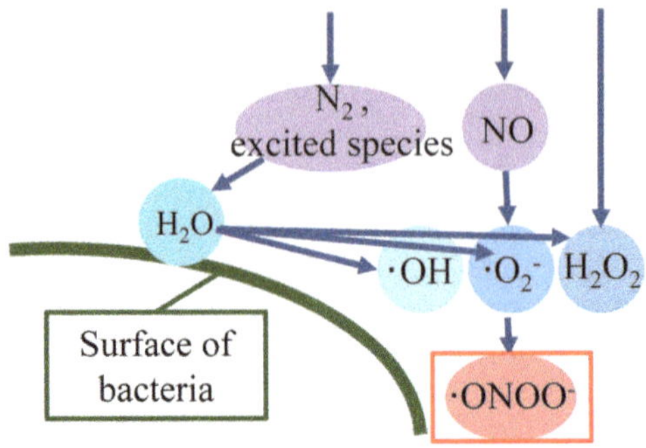

Figure 4.19 Relationship between fluorescence intensity and relative humidity (RH) for APF and HPF in Fig. 4.18. HPF does not detect significantly ·OH, indicating that the intensity of APF may be due to ONOO·⁻ and the tendency of the ONOO· ⁻ produced coincides with that of RH, which indicates the optimum RH is 0.5%.

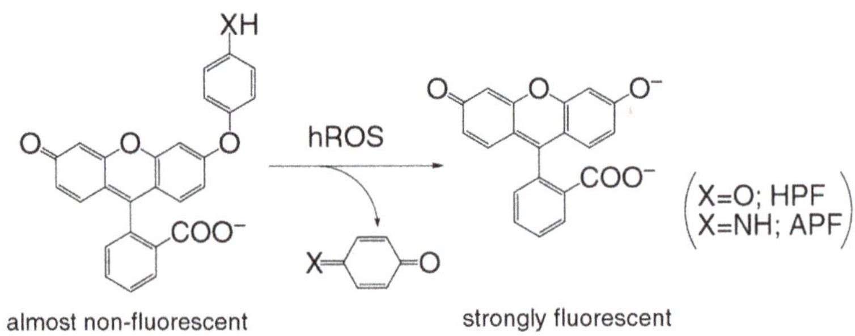

Figure 4.20 Relationship between sterilization efficiency and water vapour supply portion. Location (i) was found to be the most appropriate.

with respect to sterilization efficiency. The reason (ii) in Fig. 4.4 was inferior to the others was likely due to N metastables or OH radicals being inactivated before reaching the BI target. Location (iii) in Fig. 4.4 represents the shortest distance between the water vapour supply and the site of sterilization, whereas location (i) in Fig. 4.4 was the most remote, but the abundantly produced NO radicals are the precursors of $OONO_2\cdot^-$ (peroxynitrite anion radicals), which are the real sterilization factors described later. Measurement of NO radicals was conducted using Figs. 4.11 and 4.12. NO radical detection using CI was also reported by Shintani *et al.* (2014). Additional results supported the conclusion that water vapour supply location (i), rather than location (iii) resulted in the best sterilization efficiency.

The relationship between sterilization efficiency and relative humidity combined with several ROSs is presented in Fig. 4.21. The results indicate that sterilization efficiency coincides with the tendency of peroxynitrite anion radical ($ONOO_2\cdot^-$) formation; therefore, peroxynitrite anion radicals ($ONOO_2\cdot^-$) are thought to be the major factor of nitrogen gas plasma sterilization. Other factors such as NO radicals, H_2O_2, OH radicals or $O_2\cdot^-$ do not coincide with the% RH (Figs. 4.12–4.14). Peroxynitrite anion radicals ($ONOO_2\cdot^-$) react with tyrosine, causing nitration at the p site and with DNA bases, especially guanine, causing nitration (-NO_2) and hydroxylation (-OH), which results in transcription failure.

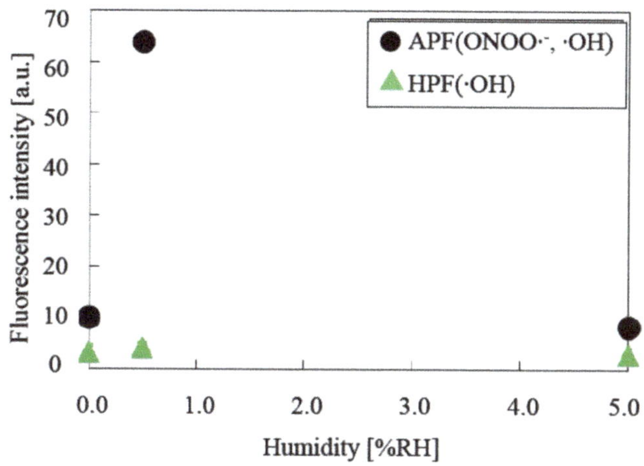

Figure 4.21 Relationship between sterilization efficiency and relative humidity (RH) for NO radical, H_2O_2 or $ONOO\cdot^-$. $ONOO\cdot^-$ coincides with sterilization efficiency.

Conclusion

The experiments reported here were conducted to identify the nitrogen gas plasma sterilization factor(s) and the appropriate sterilization conditions. By varying hotplate temperature, RH and water vapour supply location, sterilization efficiency was confirmed. In addition, SEM observation of spore surfaces, emission spectrophotometric analysis, and determination and evaluation of peroxynitrite anion radicals ($ONOO\cdot^-$) were conducted to determine which ROSs contribute to nitrogen gas plasma sterilization.

The sterilization times at 55°C, 65°C and 75°C were 240 min, 150 min, and 120 min, respectively, indicating that at higher hotplate temperatures, the sterilization periods were shorter. Increasing the temperature by 20°C reduced the sterilization period by half.

The sterilization efficiency was improved by using a combination of water vapour and nitrogen gas. Relative humidity (RH) was changed from 0.0% RH, 0.5% RH and 5% RH and the *D* values under these conditions were 10.71 min, 8.66 min and 10.07 min, respectively, indicating that the optimum RH is 0.5%. In order to identify the sterilization factors, the water vapour supply location was varied. The results indicate that the active species were relatively long-lived because the most efficient location was the most remote from the reactor.

SEM observation indicated that there was no significant difference in the appearance of control and treated spores, and no etching occurred. Treated spores seemed to have increased roughness compared with control spores, but this roughness did not always increase with increasing sterilization time, so roughness is not always an indication of sterilization. The reason has not been clarified, so no ROSs can be confirmed from SEM observation

By attaching a quartz window to the reactor, it was possible to carry out emission spectrophotometric analysis. Based on the emission spectrum at a RH of 0.5%, NO radicals, N_2 second positives and N_2^+ were detected (Fig. 4.12). In this experiment, NO radicals, which are detected at 258.55 nm in the UV-C range, increased with increasing relative humidity (Fig. 4.13). This indicates that the tendency of NO radical formation does not coincide with that of the sterilization tendency as shown in Fig. 4.21. This result indicates that NO radicals do not participate directly as a major factor in nitrogen gas plasma sterilization.

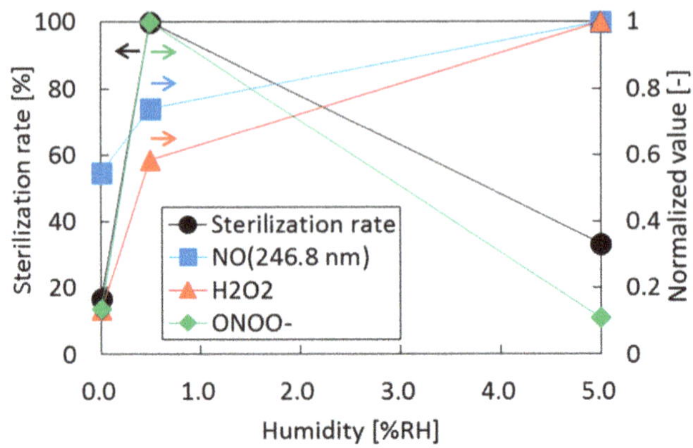

Figure 4.22 Relationship between sterilization efficiency and relative humidity (RH). 0.5% RH was found to be the optimal condition. The optimum relative humidity (RH) was determined. A RH of 0.5% was the most appropriate for sterilization. This result has been confirmed in another experiments with consistent results.

ROSs such as NO radicals, H_2O_2, OH radicals, $O_2 \cdot^-$ (superoxide anion radicals) or $ONOO_2 \cdot^-$ (peroxynitrite anion radicals) (Fig. 4.21) as well as NOx and ozone were compared for their contribution to sterilization. NOx and ozone were determined and their amounts were less than 0.6 ppm and 0.04 ppm, respectively. Since the amounts generated were so low, it can be concluded that these ROSs do not contribute to nitrogen gas plasma sterilization. The RH tendency coincided with that of $OONO_2 \cdot^-$ (Figs. 4.19 and 4.21); therefore, we conclude that $OONO_2 \cdot^-$ may be the major sterilization factor in nitrogen gas plasma sterilization.

Based on the experimental conditions for nitrogen gas plasma sterilization, the water vapour supply position was best at location (i) (furthest from the reactor; Fig. 4.4) and the humidity was optimal at 0.5% RH (Fig. 4.22). Hotplate temperature was optimal at 75°C (Fig. 4.5). Together these results indicate that higher temperature and optimum RH at 0.5% were the best when using position (i) for the water vapour supply (Figs. 4.8, 4.9 and 4.20). All results support the concluding data summarized in Fig. 4.21.

In this chapter, we introduce our recent studies on nitrogen gas plasma for remote sterilization and show the original description of $ONOO_2 \cdot^-$ (peroxynitrite anion radical). Then, we concluded that $ONOO_2 \cdot^-$ is the major factor in nitrogen gas plasma sterilization. In contrast to $ONOO_2 \cdot^-$ (Figs. 4.19 and 4.21), other ROSs do not have identical tendencies with respect to RH (Figs. 4.13, 4.14 and 4.22) and only $ONOO_2 \cdot^-$ presents an identical tendency to the RH. Therefore, $ONOO_2 \cdot^-$ can be defined as the major sterilization factor in nitrogen gas plasma sterilization.

Acknowledgement

This chapter is reproduced from a previous article (Shintani *et al.*, 2015a) written by the same corresponding author (Hideharu Shintani) and distributed under the terms of the Creative Commons Attribution License (CCAL). It should be noted that CCAL permits unrestricted use, distribution and reproduction in any medium.

References

Deng, X., Shi, J., and Kong, M.G. (2006). Physical mechanisms of inactivation of *Bacillus subtilis* spore using cold atmospheric plasma. IEEE Transactions Plasma Sci. *34*, 1310–1316.

Friedman, A and Friedman, G. (eds). (2013). Effects of water and its amount. In Plasma Medicine, pp. 281–282, Wiley Ltd, UK.

ISO 11138-1 (2009). Sterilization of health care products – Biological indicators – Part 1: General requirements.

ISO 14161 (2009). Sterilization of health care products – Biological indicators – Guidance for the selection, use and interpretation of results.

ISO 11737–1 (2010). Sterilization of medical devices – Microbiological methods – Part 1: Determination of a population of microorganisms on products.

Li, J., Sakai, N., Watanabe, M., Hotta, E., and Wachi, M. (2013). Study on plasma agent effect of a direct-current atmospheric pressure oxygen-plasma jet on inactivation of E.coli using bacterial mutant. IEEE Transactions Plasma Sci. *41*, 935–941.

Maeda, K., Toyokawa, Y., Shimizu, N., Imanishi, Y., and Sakudo, A. (2015). Inactivation of Salmonella by nitrogen gas plasma generated by static induction thyristor as a pulsed power supply. Food Control *52*, 54–59.

Novo, E., and Parola, M. (2008). Redox mechanisms in hepatic chronic wound healing and fibrogenesis. Fibrogenesis and Tissue Repair *1*:5, pp.1–58.

Sakudo, A., Shimizu, N., Imanishi, Y., and Ikuta, K. (2013). N_2 gas plasma inactivate influenza virus by inducing changes in viral surface morphology, protein, and genomic RNA. BioMed Res. International, Article ID 694269, 7 pages.

Sakudo, A., Misawa, T., Shimizu, N., and Imanishi, Y. (2014). N_2 gas plasma inactivates infliuenza virus mediated by oxidative stress. Frontiers in Bioscience *6*, 69–79.

Sakudo, A., and Shintani, H. (eds.) (2012). Sterilization and Disinfection by Plasma, Sterilization Mechanisms, Biological and Medical Applications. NOVA Science Publishers, Inc. New York.

Setsukinai, K., Urano, Y., Kakinuma, K., Majima, H.J., and Nagano, T. (2003). Development of novel fluorescence probes that can reliably detect reactive oxygen species and distinguish specific species. J. Biol. Chem. *278*, 3170–3175.

Shimizu, N. (2010). The study of High-speed Switching Characteristics of Static Induction Thyristor and its Applications, Tohoku University, Dr. dissertation.

Shintani, H. (1995). The relative safety of bgamma-ray, autoclave, and ethylene oxide gas sterilization of thermosetting polyurethane. Biomed. Instrument. Technol. *29*, 513–519.

Shintani, H. (2014). Comparison on safety concern among gamma-ray, autoclave and ethylene oxide gas sterilizations of thermosetting polyurethane. Pharmaceut. Reg Affairs 3, e139.

Shintani, H. (2015). Sterilization mechanisms of nitrogen gas plasma exposure. Int. J. BioAnalytical Methods BioEquivalence Studies 2 (1e).

Shintani, H., Shimizu, N., Imanishi, Y., Sekiya, T., Tamazawa, K., Taniguchi, A., and Kido, N. (2007). Inactivation of microorganisms and endotoxins by low temperature nitrogen gas plasma exposure. Biocontrol Sci. *12*, 131–143.

Shintani, H., Sakudo, A., Burke, P., and McDonnell, G. (2010). Gas plasma sterilization of microorganisms and mechanisms of action. Experimental Therapeutic Medicine, *1*, 7381–7738.

Shintani, H., Saito, T., and Nakamura, K. (2014). Chemical indicator and ink constitute to detect several radicals, metastables and ions produced by gas plasma sterilization. Japan Patent. 233387.

Shintani, H., Shimizu, N., Imanishi, Y., Sakudo, A., Takuya, U., and Hotta, E. (2015). Efficiency of atmospheric pressure nitrogen gas remote plasma sterilization and the clarification of sterilization major factors. Int. J. Clin. Pharmacol. Toxicol. *4*, 150–160.

Tamazawa, K., Shintani, H., Tamazawa, Y., and Shimauchi, H. (2015). Sterilization effect of wet oxygen plasma on bubbling method. Biocontrol Sci. *20*, in press.

Uyama, T. (2015). Sterilization efficiency of atmospheric nitrogen remote plasma using pulsed discharge and study for sterilization factors. Tokyo Institute Science and Technology, Masters Thesis.

Current Progress in the Inactivation of Endotoxin and Lipid A by Exposure to Nitrogen Gas Plasma

5

Hideharu Shintani

Abstract

Nitrogen gas plasma treatment has sporicidal activity as well as the ability to inactivate endotoxins and lipid A. The mechanism of nitrogen gas plasma sterilization may include synergistic effects involving free radicals (e.g. OH, NO or OONO radicals) and metastable species (i.e. metastable states of N_2 or O_2). Exposure to nitrogen gas plasma caused no discernible deterioration in the functional compatibility of various materials under investigation. Based on these findings, nitrogen gas plasma sterilization is a promising method for the sterilization of medical devices.

Introduction

Recent searches of the primary literature and internet using the keywords endotoxin, gas plasma, inactivation or depyrogenation resulted in only a few matching articles, indicating that studies of endotoxin inactivation/depyrogenation using gas plasma are very limited (Shintani *et al.*, 2007, 2010; Fujimori and Arkawa, 1998; Vetten *et al.*, 2014; Tessarolo *et al.*, 2006; Keudell *et al.*, 2010). However, the degree of endotoxin inactivation by Sterad® was reported to be approximately a 1-log reduction (Hosobuchi and Tanamoto, 1999; Tamazawa and Hosobuchi, 2004). Our method of endotoxin inactivation involved the use of nitrogen gas plasma exposure. In this study, nitrogen gas plasma treatment resulted in an endotoxin reduction of more than a 5 logs in 30 min (Fig. 5.1; Shintani *et al.*, 2007). These experimental findings indicate that the performance characteristics of nitrogen gas plasma are superior to those of Sterad®, which has been reported to use hydrogen peroxide gas plasma. However, Sterad® is not a true hydrogen peroxide gas plasma sterilizer due to the large chamber size (100–150l) of Sterad^R (Shintani *et al.*, 2007; Shintani, 2012).

Inactivation of endotoxin

As shown in Fig. 5.1, at higher temperatures, a greater degree of endotoxin inactivation was obtained. Since the degree of inactivation was dependent on temperature, the inactivation follows the Arrhenius equation. Inactivation of endotoxins was confirmed from the results of *Limulus* ES-II test Wako and Toxinometer ET-2000/J System, Wako. Endotoxins were from LPS (lipopolysaccharide) of *E. coli* 0111 (Fig. 5.2). LPS was evenly inoculated onto

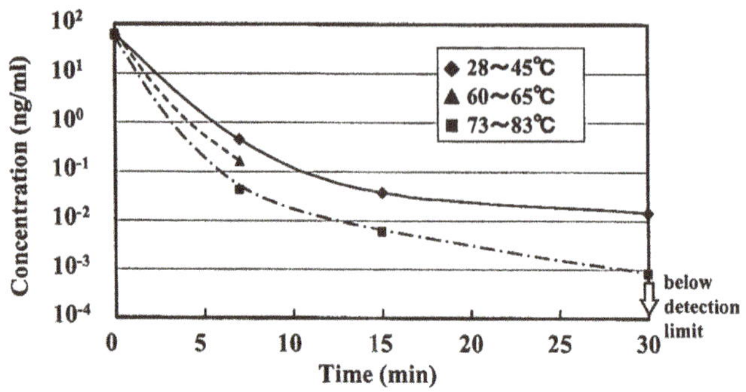

Figure 5.1 Temperature dependence of endotoxin inactivation by low pressure (LP). LP conditions: input power, 84 W; frequency, 2.5 kHz; temperature, 28–73°C, vacuum, 45 kPa; gap, 40 mm; N_2 flow rate, 15 l/min; voltage,19 kV. Reproduced from Shintani (2015).

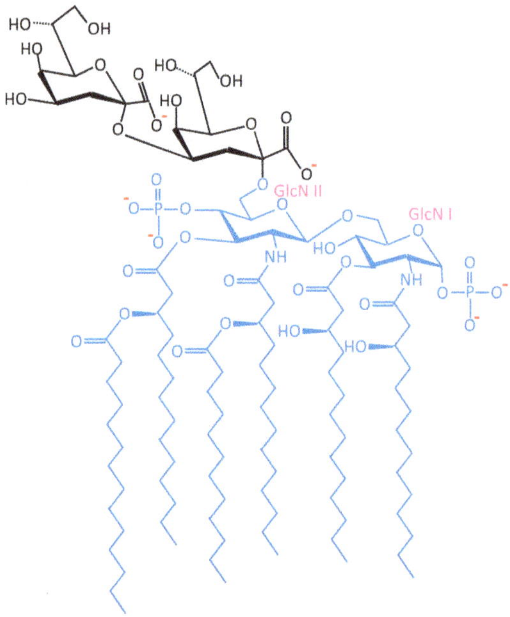

Figure 5.2 The basic lipopolysaccharide of *E. coli*, incorporating lipid A (lower portion of the structure). Reproduced from https://en.wikipedia.org/wiki/Lipopolysaccharide

modified PS (polystyrene) at the nanomolar level and thereafter exposed to nitrogen gas plasma. After inactivation of endotoxins by nitrogen gas plasma, no deformation or deterioration of the PS of the carrier material was observed, indicating that the PS maintained its material functionality (Tables 5.1 and 5.2) (Shintani *et al.*, 2007; Sakudo and Onodera, 2013; Williams, 2007; McDonnell, 2013).

Conventional methods of endotoxin inactivation are by dry heat treatment at 250°C for more than half an hour (i.e. 45 min) or gamma-ray irradiation at 100 kGy after 1N NaClO immersion (Tamazawa and Hosobuchi, 2004). However, material/functional compatibility

Table 5.1 XPS result for PS (polystyrene)

Sample	C (%)	O (%)	N (%)	Si (%)
PS (before treatment)	98.7	1.3		
PS (nitrogen gas plasma treatment)	82.9	14.2	2.2	0.7

Condition of nitrogen gas plasma treatment: 1/2 atmospheric pressure, 60°C, 6 min.
Reproduced from Shintani *et al.* (2007).

Table 5.2 Analysis of the exhaust gas from PS (polystylene) treated with nitrogen gas plasma

Disposition	Before or after treatment with PS	CO[1] (v/v, ppm)	NOx[2] (v/v, ppm)	HCN[3] (v/v, ppm)	O_3[4] (v/v, ppm)	N_2O[5] (v/v, ppm)
Low pressure	Before	<2	<0.5	ND	ND	ND
Low pressure	After	3.9	1.1	<0.1	<0.05	2.6

[1]UV-absorbance spectroscopy.
[2]Chemical luminescence method.
[3]Piasoron light absorption method.
[4]Ozone detector.
[5]GC-MS.
ND, not detected.
Reproduced from Shintani *et al.* (2007).

may fail to be maintained with these procedures. In contrast, treatment with nitrogen gas plasma resulted in more than a 5 log reduction of endotoxin in ~20–30 min while simultaneously maintaining material/functional compatibility. These findings indicate that nitrogen gas plasma treatment is superior to the conventional procedures of endotoxin inactivation (Sehulster, 2012).

Mechanism of endotoxin inactivation using lipid A

The mechanism of endotoxin inactivation was studied by using lipid A and HPLC-MS-MS (high-performance liquid chromatography–mass spectrometry–mass spectrometry). Synthesized lipid A from Peptide Institute, Inc. (Osaka, Japan) was used to examine the endotoxin inactivation mechanism. This is because lipid A contains the active site of endotoxin (Brandenburg *et al.*, 2001; Tzeng *et al.*, 2002; Hung, *et al.*, 2014; Chang *et al.*, 2014; Poon, 2011; Aussel *et al.*, 2000; Zarrouk *et al.*, 1997; Brandtzaeg *et al.*, 1989; Kedia *et al.*, 2014). The chemical structures of endotoxin and lipid A from *E. coli* are presented in Figs. 5.2 and 5.3, respectively. Combined chemical structure is shown in Fig. 5.4.

In Fig. 5.5, the M-l peak (negative peak) was observed to be 1797 m/z and represents the mother peak of lipid A (Chalabaev *et al.*, 2014). In addition, peaks at 209, 247, 297, 311, 325, 339, 415, 429, 502, 603, 632, 645, 660, 673 and 770 m/z were observed in the non-treated sample, which was dissolved in DMSO (Fig. 5.5a). However, in the sample treated by nitrogen gas plasma (Fig. 5.5b), these peaks were absent, and, moreover, lower molecular weight peaks decreased significantly. Only very small MS peaks were present in the MS spectrum of the plasma-treated sample. This indicates that cleavage at several positions in lipid A, such as the acid amide bond or ester bond, can be speculated to occur as a result of the exposure to nitrogen gas plasma. As shown in Fig. 5.5c, the mother peak can

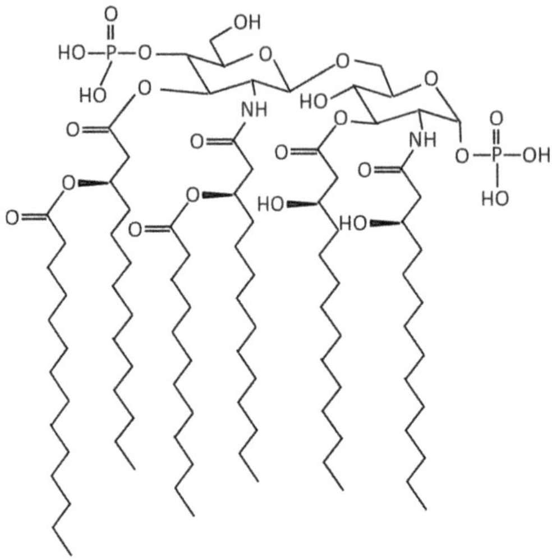

Figure 5.3 Chemical structure of lipid A. This lipid A is from *E. coli*. Reproduced from https://en.wikipedia.org/wiki/Lipid_A

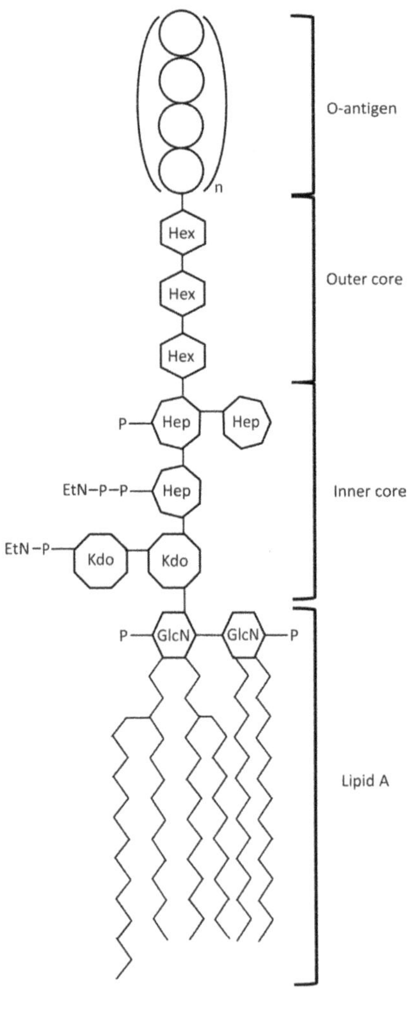

Figure 5.4 Chemical structure of endotoxin. Cited from http://www.science.mcmaster.ca/biochem/faculty/bishop/publications/pdf/2005_LPS_Chapter.pdf

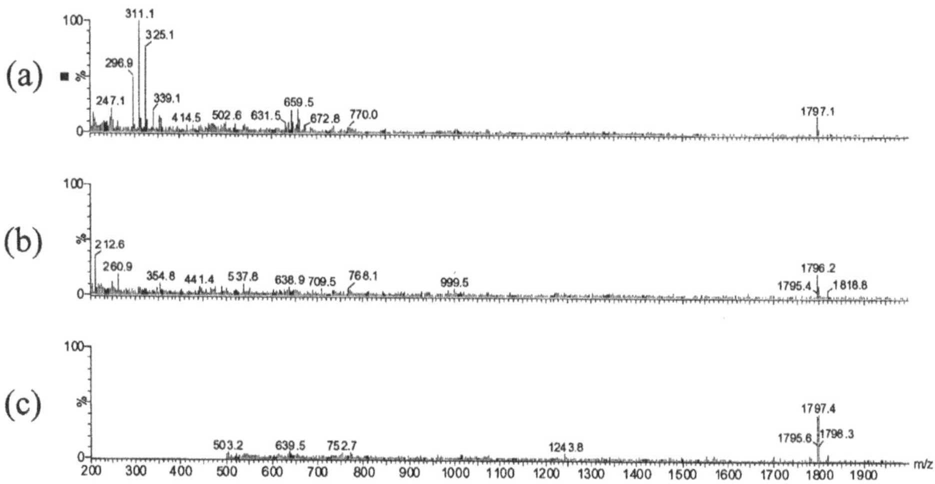

Figure 5.5 MS spectrum of lipid A before and after nitrogen gas plasma exposure. (a) is before exposure, (b) is after exposure and (c) is DMSO injection. LC-MS from Waters was used. LC was Waters 2695 and MS was Waters Quattro Premier^{R} XE.LC column was Golf Pack^{R}. HR 3.5µm 2.1 × 150 mm (polymer base C18 column), Mobile phase was acetonitrile, Flow rate was 0.2 ml/min, Injection volume was 5 µl. MS mode. ESI negative mode was used for ionization. Capillary voltage was 4.0 kV, solvent degassing rate was 900 l/h, cone gas rate was 50 l/h and ion source heater was 120°C. MS data scanning conditions were as follows: MS scanning range was 200–2000 Da, scan time was 0.5 s and cone voltage was 150 V. Reproduced from Shintani (2015).

be observed in the DMSO solvent alone injection, which means that the mother peak may remain in the injection port and/or LC column as an artefact. Thus, due to the appearance of this artefact, we unfortunately cannot compare the peak height of the mother peak before and after nitrogen gas plasma exposure. However, it is clear that the peaks of around 200 and 700 m/z in lipid A decreased as a result of nitrogen gas plasma treatment (Fig. 5.5a and b).

The mass of peaks at around 200–300 m/z may be single or double chains of C14 fatty acids linked with ester bonds, and peaks at ~600–700 m/z may be multiple chains of C14 fatty acids linked with ester bonds. The exact site(s) in lipid A that were cleaved by nitrogen gas plasma exposure were not identified (Peeples and Anderson, 1985a,b), but we can speculate that bonds with lower bonding energy such as ester or acid amide bonds may have been cleaved (C(=O)-O, C(=O)-NH; Table 5.3).

Degradation of lipid A by nitrogen gas plasma exposure was observed by atomic force microscopy (AFM) (Fig. 5.6) and X-ray photoelectron spectroscopy (XPS) (Fig. 5.7). Only a trace of lipid A that was inoculated onto glass remained after nitrogen gas plasma exposure based on AFM analysis. Moreover, XPS results were identical (Table 5.1), and showed an increase in organic nitrogen due to increased binding energy at organic nitrogen peaks of around 400 to 404 eV after treatment. Thus, at the nanomolar level, lipid A was almost completely degraded by nitrogen gas plasma exposure (Fig. 5.6).

Endotoxin inactivation is essential in implanted medical devices to avoid pyrogen shock in patients, which may even cause death in the worst case. In that respect, the attainment of endotoxin inactivation at a reduction of more than 5 logs by nitrogen gas plasma exposure in 30 min has important potential for application to medical devices prior to implantation into

Table 5.3 Average bonding energy (kJ/mol)

Chemical bonding	Bonding energy (kJ/mol)
H-H	436
C-C	344
C=C	615
C≡C	812
O-O	143
C-O	350
C=O	725
C-H	415
N-H	391
O-H	463

Reprduced from Shintani *et al.*, 2007).

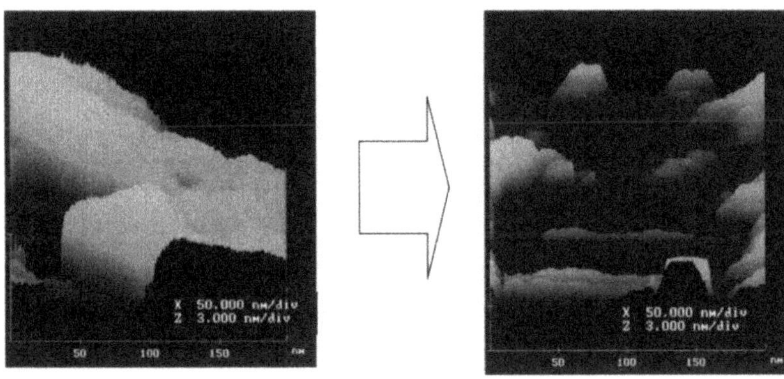

Figure 5.6 AFM analysis of LP plasma-treated Lipid A (1.25 µg), which was inoculated onto glass. LP, lower pressure. Reproduced from Shintani (2015).

the human body. Adoption of this process would have significant benefit to patients requiring medical device implantation by ensuring endotoxin-free devices to prevent disease conditions associated with contaminating endotoxins such as pyrogen shock. Currently, the relevant authorities require less than 10 EU (endotoxin unit)/ml while maintaining material/functional compatibility (Tables 5.1 and 5.2), and this requirement can easily be achieved with our method.

Since we have confirmed the destruction of spore-type microorganisms and endotoxin inactivation by nitrogen gas plasma exposure, we will conduct experiments on prion inactivation by nitrogen gas plasma exposure without carrier material deterioration (Shintani *et al.*, 2007, Shintani, 2012). According to our preliminary experiments, normal prions were totally destroyed as had been the case with spores and endotoxins. Based on this result, we speculate that abnormal prions that cause Creutzfeldt–Jakob disease to humans would also be destroyed by nitrogen gas plasma exposure without deterioration of the support material, which would be extremely useful for preventing transmission of prion-mediated diseases.

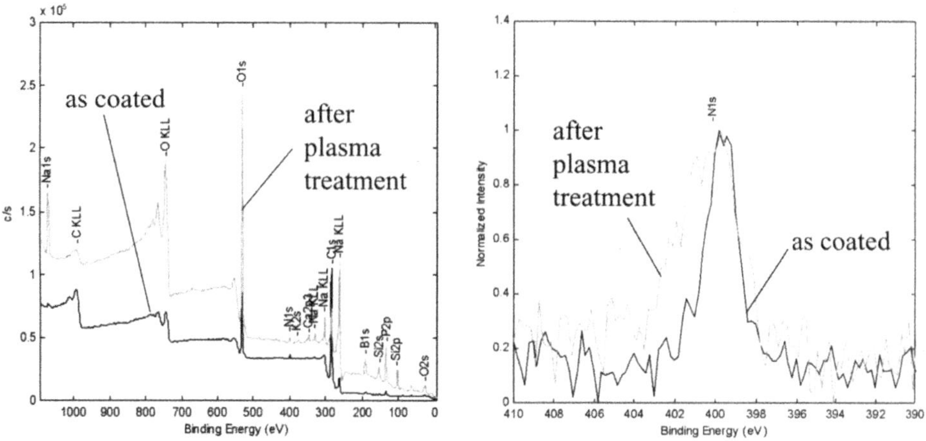

Figure 5.7 XPS analysis of LP plasma treated lipid A (1.25 μg), which was inoculated onto glass. LP, lower pressure. Reproduced from Shintani (2015).

We speculate that abnormal prions can be successfully destroyed while maintaining material functionality because abnormal and normal prions have identical primary amino acid sequences and differ only in regard to their higher dimension of structure. Abnormal prions have more beta sheet structure than normal prions. We already confirmed that myoglobin, which is enriched in beta sheet structure, is degraded by nitrogen gas plasma exposure while maintaining material/functional compatibility. In addition, in future studies we will conduct experiments to validate the sterilization of the interior of endoscopes, expensive artificial heart valves, expensive surgery devices used for brain or nerve surgeries, etc., by nitrogen gas plasma. Successful use of this technology for these applications would be beneficial for patients requiring these devices to avoid iatrogenic diseases caused at health care facilities.

Conclusion

Endotoxin inactivation can be successfully achieved by exposure to nitrogen gas plasma for 30 min. After this treatment, more than a 5 log reduction of endotoxin can be confirmed. The authorities require less than 10 EU/ml of endotoxin together with material/functional compatibility, and our method can successfully achieve these requirements without any difficulties.

Lipid A is an active site of endotoxin. Lipid A degradation after nitrogen gas plasma exposure was observed using high-performance liquid chromatography–mass spectrometry. From the HPLC-MS-MS data, fragmentation and degradation of lipid A by nitrogen gas plasma was confirmed. Normal prion and myoglobin degradation was also confirmed.

Acknowledgement

This chapter is reproduced from a previous article (Hideharu Shintani, Pharmaceut Reg Affairs, 2015, 4: 129; doi: 10.4172/2167-7689.1000129) written by the same corresponding author (Hideharu Shintani) and distributed under the terms of the Creative Commons

References

Aussell, L., Brisson, J.R., Perry, M.B., and Caroff, M. (2000). Structure of the lipid A of *Nordetella hinziii* ATCC 51730. Rapid Commun. Mass Spectrometry *14*, 595–599.

Brandenburg, K., Jurgens, G., Muller, M., Fukuoka, S., and Koch, M.H. (2001). Biophysical characterization of lipopolysaccaride and lipid A inactivation by lactoferrin. Biol. Chem. *382*, 1215–1225.

Brandtzaeg, P., Kieruf, P., Gaustad, P., Skulberg, A., Bruun, J.N., Halvorsen, S., and Serensen, E. (1989). Plasma endotoxin as a predictor of multiple organ failure and death in systemic meningococcal disease. J. Infect. Dis. *159*, 195–204.

Chalabaev, S., Chauhan, A., Novikov, A., Iyer, P., Szczesny, M., Beloin, C., Caroff, M., and Ghigo, J.-M. (2014). Biofilms formed by gram-negative bacteria undergo increased lipid A palmitoylation, enhancing *in vivo* survival. Mbio.asm.org. *5*, e01116–14.

Chang, H.-W., Hsu, C.-C., Abmed, M., Liu, S.Y., Fang, Y., Seog, J., Oehrlein, G.S., and Graves, D.B. (2014). Plasma flux-dependent lipid A deactivation. J. Physics. Appl. Phys. *47*, 224015 doi:10.1088/0022-3727/47/22/224015 (http://dx.doi.org/10.1088/0022-3727/47/22/224015)

Fujimori, K., and Arakawa, M. (1998). Clinical factors that affect blood gases in non-smoking women with chronic liver disease, Repair Med. *92*, 57–62.

Hosobuchi, K., and Tanamoto, K. (1999). Tokyo Metropolitan Industrial Institute Research Report, *2*, 126–129.

Kedia, A., Prakash, B., Mishra, P.K., and Dubey, N.K. (2014). Antifungal and antiaflatoxigenic properties of Cuminum cyminum (L) seed essential oil and its efficiency as a preservative in stored commodities. Int. J. Food Microbiol. 2014 Jan 3;168–169;1–7.doi;10.1016

Keudell, A., Awakowicz, P., Benedikt, J., Raballand, V., Yanguas-Gil, A., Opretzka, J., Floetgen, C., Reuter, R., Byelykh, L., Halfmann, H., *et al.* (2010). Inactivation of bacteria and biomolecules by low-pressure plasma discharges. Plasma Process Polym. *7*, 1–26.

McDonnell, G.E. (2013). Gas plasma sterilization in Principles and Practice of Disinfection, Prerservation and Sterilization, 5th edn. Fraise, A., Maillard, J-V and Sattar, S.A. eds. (Wiley-Blackwell), Oxford, UK, pp. 333–342.

Peeples, E.E., and Anderson, N.R. (1985a). Microwave coupled plasma sterilization and depyrogenation I. Systems characteristics. J. Parent. Sci. Technol. *39*, 2–8.

Peeples, E.E., and Anderson, N.R. (1985b). Microwave coupled plasma sterilization and depyrogenation II. Mechanism of action. J. Parent. Sci. Technol. *39*, 9–15.

Poon, A. (2011). Endotoxin removal by radio frequency gas plasma (glow discharge), Dissertation & Theses-Gradworks, MS degree (2011). 127 pages,1500498. State University of New York at Buffalo.

Sakudo, A., and Onodera, T. (2013). Prions: Current Progress in Advanced Research. (Horizon Scientific Press). pp. 59–61.

Sehulster, L.M. (2012). Prions and endotoxins: reprocessing strategies for reusable medical devices. In Sterilization of Biomaterials and Medical devices. Lerouge, S., and Simmons, A., eds. (Woodhead Publishing), Oxford, pp. 261–309.

Shintani, H. (2012). Inactivation of prion and endotoxin by nitrogen gas plasma exposure, Pharmaceut. Anal. Acta 3, 1000177.

Shintani, H. (2015). Inactivation of endotoxin and lipid A by nitrogen gas plasma exposure. Pharmaceut. Reg. Affairs *4*, 129.

Shintani, H., Shimizu, N., Imanishi, Y., Sekiya, T., Tamazawa, K., Taniguchi, A., and Kido, N. (2007). Inactivation of microorganisms and endotoxins by low temperature nitrogen gas plasma exposure, Biocontrol Sci. *12*, 131–143.

Shintani, H., Sakudo, A., Burke, P., and McDonnell, G.E. (2010). Gas plasma sterilization of microorganisms and mechanisms of action. Experment. Therapeutic Med. *1*, 731–738.

Tamazawa, K., and Hosobuchi, K. (2004). Development of endotoxin inactivation by oxygen plasma. Jap. J. Med. Instrument. *74*, 285–291.

Tessarolo, F., Caola, I., Nollo, G., Antolini, R., Guarrera, G.M., and Caciagli, P. (2006). Efficiency in endotoxin removal by a reprocessing protocol for electrophysiology catheters based on hydrogen peroxide plasma sterilization. Int. J. Hyg. Environ. Health *209*, 557–565.

Tzeng, Y.L., Datta, A., Kolli, V.K., Carlson, R.W., and Stephens, D.S. (2002). Endotoxin of *Neisseria meningitidis* composed only of intact lipid A: inactivation pf the meningococcal 3-deoxy-D-manno-octulosonic acid transferase. J. Bacteriol. *184*, 2379–2388.

Vetten, M.A., Yah, C.S., Singh, T., and Gulumian, M. (2014). Challenges facing sterilization and depyrogenation of nanoparticles: Effects on structural stability and biomedical applications, Nanomedicine *1000*, 1391–1399.

Williams, K. (2007). Depyrogenation validation, pyroburden, and endotoxin removal In Endotoxins, Pyrogens, LAL, Testing and Depyrogenation, 3rd, Williams, K.L., ed. (Informa), pp. 301–303.

Zarrouk, H., Karibian, D., Bodies, S., Perry, M.B., Richards, J.C., and Caroff, M. (1997). Structural characterization of the lipids A of three *Bordetella bronchiseptica* strains; variability of fatty acid substitution. J. Bacteriol. *179*, 3756–3760.

Current Progress in Advanced Research into Tetrodotoxin Inactivation by Gas Plasmas

6

Toshihiro Takamatsu, Hidekazu Miyahara, Takeshi Azuma and Akitoshi Okino

Abstract

This study treated tetrodotoxin (TTX) solution with non-thermal multi-gas plasma and analysed its decomposition by liquid chromatography coupled with electrospray time-of-flight mass spectrometry. The TTX signal in the mass spectrum was reduced to different levels by plasma irradiations using various gases. Nitrogen plasma exhibited the optimal capability for TTX decomposition, followed by oxygen, argon, and carbon dioxide plasmas. The TTX concentration decreased 100-fold after 10 min treatment with nitrogen plasma. To better understand the TTX degradation process, plasmas of five different gases were generated by a multi-gas plasma jet. The OH radicals and ozone molecules formed at the solution interface were then measured by electron spin resonance and photometry. The largest amount of ozone (64 μM at 15 s) and OH radical (130 μM at 30 s) were generated by oxygen and nitrogen plasma, respectively. We concluded that the generated reactive oxygen species such as OH radicals and ozone contribute to TTX degradation.

Introduction

In recent years, atmospheric plasma has been studied for applications in diverse fields such as semiconductor processes (Kumagai *et al.*, 2007), decomposition of harmful gases (Tamura *et al.*, 2011) and substances (Watanabe and Tsuru, 2008), and elemental analysis (Shigeta *et al.*, 2013). An atmospheric plasma source can generate high-density plasma and enables a continuous plasma treatment without requiring a vacuum chamber and an exhaust system. Therefore, it can provide high-efficiency, high-speed treatments, which are desired in various industrial fields. In particular, atmospheric non-thermal plasma can be generated at low temperature (room temperature to 100°C). Consequently, atmospheric plasma has attracted great attention. Non-thermal plasmas using highly reactive species are reported to sterilize *E. coli* (Shimizu *et al.*, 2008) and spore bacteria such as *B. subtilis* (Takamatsu *et al.*, 2011). They can also hydrophilize surfaces (Takamatsu *et al.*, 2013) and reduce the oxidation of film surfaces (Nakashima *et al.*, 2012). These effects depend on the active species generated by the plasma. For example, during organic material hydrophilization by non-thermal plasma, molecular chains such as C-C, C-H, and C-O are thought to be removed by reactive species reacting on the surface, generating hydrophilic functional groups such as carboxyl and carbonyl (Xie *et al.*, 2011). In addition, because the generated plasma is touchable and biologically friendly, it has been investigated as a body disinfectant (Isbary *et al.*,

2012) and a food sterilizer (Ragni *et al.*, 2010). Considering these advantages, we proposed that atmospheric non-thermal plasma might also decompose toxic substances with organic bonds, which resist decomposition by both heat and chemical reaction. Among these substances is tetrodotoxin (TTX), more commonly known as puffer fish toxin. Ingesting just 1–2 mg TTX can be fatal, even leading to death.

Although TTX is stable at high temperature (> 300°C), it can be decomposed by strong acid or alkaline antagonistic agents, and TTX poisoning generally requires special therapy. In this study, TTX solutions were treated with multi-gas plasmas, and their decomposition was analysed by liquid chromatography coupled with electrospray time-of-flight mass spectrometry (LC-ESI–TOF-MS). In addition, the amount of reactive species generated by the plasma was identified by electron spin resonance (ESR) and photometric measurements. By quantifying the reactive species, we can better understand the reaction process of TTX decomposition.

Atmospheric multi-gas plasma jet

Fig. 6.1 is a photograph of our plasma jet source. This source generates stable atmospheric-pressure plasmas of various gases; helium, argon, oxygen, neon, nitrogen, carbon dioxide, air, and mixtures of these gases at low gas temperature (<57°C) and approximately 10 W power (Takamatsu *et al.*, 2013). The source is easily operated by virtue of its small body (length 83 mm; weight 160 g). The body is grounded and the interior high-voltage electrode is connected to a power supply (Plasma Concept Tokyo, Inc.) operating at 16 kHz and 9 kV. The discharge gap, defined as the distance between the high-voltage and grounded electrodes, is fixed at 1.5 mm. The electrodes are composed of aluminium, and each electrode is 12.6 mm^2 in area. The generated plasma flows out through a 1-mm-diameter hole.

TTX degradation

The chemical structure of TTX is shown in Fig. 6.2. A 10 ppm TTX (Cellular biology grade, Wako Pure Chemical Industries, Ltd., Osaka, Japan) solution was prepared in ultrapure water, and 1 ml of this solution was added to a 1.5-mL micro-tube. The sample was subjected

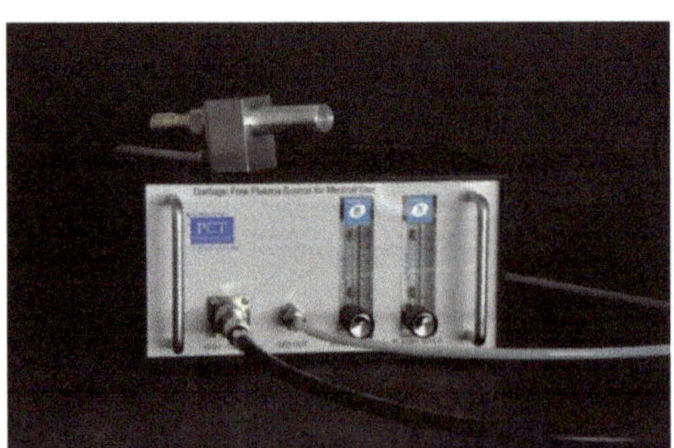

Figure 6.1 Damage-free multi-gas plasma jet, the power supply is operated at 16 kHz and 9 kV.

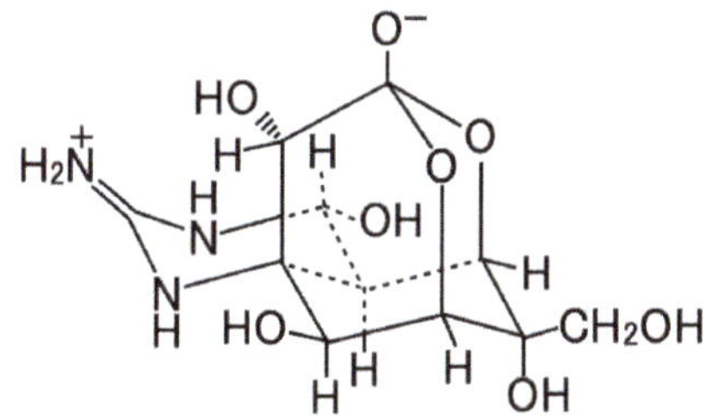

Figure 6.2 Chemical structure of tetrodotoxin (TTX).

to a multi-gas plasma jet built in the laboratory (Takamatsu *et al.*, 2013). The gas flow rate was maintained at 2 l/min, and the treated TTX solution was placed 15 mm from the plasma source outlet, as illustrated in Fig. 6.3. After the plasma treatment, the solutions were analysed by LC-ESI-TOF-MS (micrOTOFII, Bruker Co., MA, USA) under the analytical conditions listed in Table 6.1. The mass spectra of solutions treated with oxygen and nitrogen plasmas for various exposure times (1, 2, 5 and 10 min) were investigated, and the residual quantity of TTX was measured from the TTX standard curve. The mass spectrum intensity of TTX solution exposed to multi-gas plasmas for 5 min was also investigated. The gas species were argon, mock air (N_2:O_2 = 8:2), nitrogen, oxygen and carbon dioxide.

TTX can be measured at the m/z of the proton addition molecule $[M+H]^+$ (m/z = 320.1) (Tsai *et al.*, 2006); therefore, the spectral intensity assays the decomposition degree of TTX. Fig. 6.4 shows the mass spectra of TTX solutions treated with nitrogen and oxygen plasmas for different exposure times. After both plasma treatments, besides the $[M+H]^+$ ion at m/z = 320.1 and $[M+OH]^+$ ion at m/z = 336.1, additional peaks appeared at m/z = 318.1, suggesting that TTX was either oxygenated or decomposed by plasma exposure. In

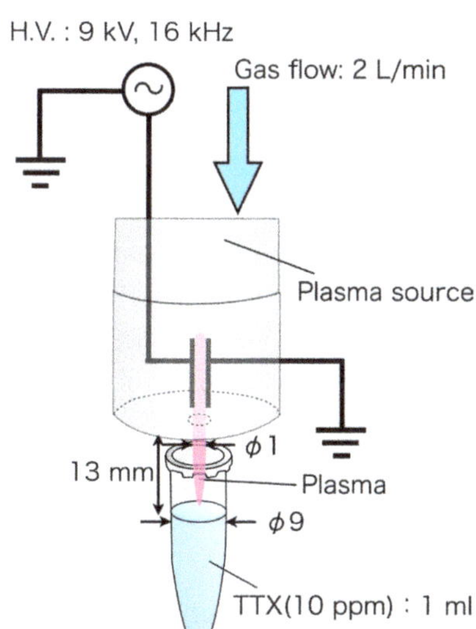

Figure 6.3 Experimental setup; Gas flow rate is 2 L/min, TTX concentration is 10 ppm, liquid volume is 1 ml and treatment distance is 13 mm.

Table 6.1 ESI-TOF-MS parameters

Scan range (*m/z*)	50–700
Ion polarity	Positive
Set capillary (V)	4500
Set end plate offset (V)	–500
Set nebulizer (bar)	1.8
Set dry gas (l/min)	10
Set divert valve	Waste
Flow rate (ml/min)	0.2
Mobile phase	50% acetonitrile
Injection volume (µl)	30

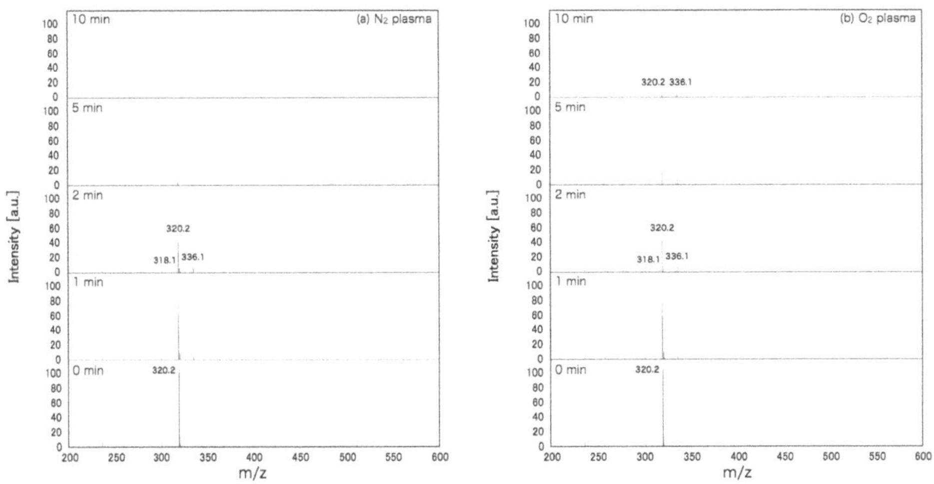

Figure 6.4 Mass spectra after (a) N_2 and (b) O_2 plasma treatment (gas flow rate = 2 l/min).

addition, the mass spectra intensity decreased as the plasma exposure time increased, indicating that prolonging the plasma exposure time more effectively degraded the TTX. The TTX concentration in samples exposed to oxygen and nitrogen plasmas for various times was determined from a TTX standard curve. As shown in Fig. 6.5, a 10-min nitrogen plasma treatment reduced the TTX concentration from 10 ppm to 0.1 ppm.

The spectrum intensities of TTX solutions treated by different gas plasmas are shown in Fig. 6.6. This result demonstrates that nitrogen plasma most effectively degrades TTX, followed by oxygen, argon, carbon dioxide, and air plasmas.

OH radical measurement of various gas plasmas using ESR

Each reactive species reacts with individual spin-trapping agents, and the spin adducts can be identified by ESR measurements. The spin-trapping agent used in this study was 5,5-dimethyl-1-pyrroline-*n*-oxide (DMPO), which detects OH radicals (Kohno *et al.*, 1991). The

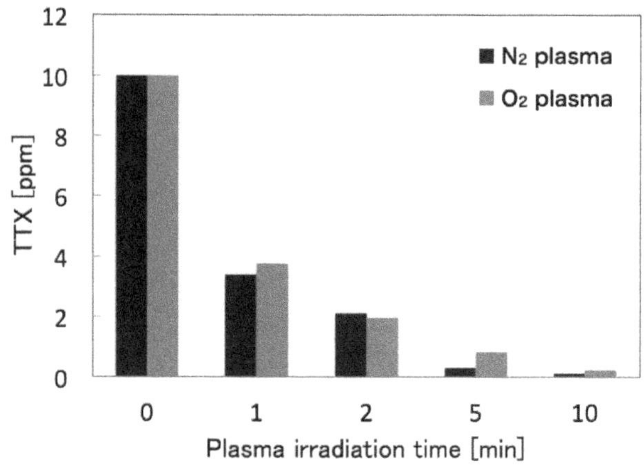

Figure 6.5 TTX concentration after treatment with N_2 and O_2 plasma for different times (gas flow rate = 2 l/min).

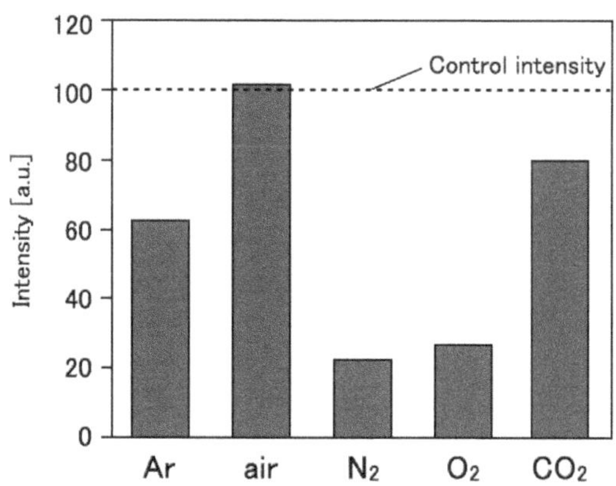

Figure 6.6 Intensity of TTX spectrum after treatment with different gas plasmas (plasma treatment time = 5 min; gas flow rate = 2 l/min).

DMPO was dissolved in phosphate-buffered saline (–) (pH 7.5) at a fixed concentration of 200 mM. The ESR settings were microwave frequency = 9.424818 GHz, sweep time = 2 min, modulation frequency = 100 kHz, magnetic field = 335.5 ± 5 mT, modulation width = 0.1 mT, and time constant = 0.1 s.

Each reactive species generated by plasmas of argon, mock air, nitrogen, oxygen, and carbon dioxide was investigated. Sample solutions of 200 μl were placed with their liquid surfaces 6 mm distant from the plasma source outlet and subjected to 30 s plasma treatment. The amount of each reactive species was calibrated with spin adducts of 2,2,6,6-tetramethyl-piperidine 1-oxyl, whose radical amount was known.

As shown in Fig. 6.7, the largest amount of OH radical of 130 μM was generated by nitrogen plasma. The OH radical produced by argon, oxygen, and carbon dioxide plasmas amounted to 84, 32, and 28 μM, respectively.

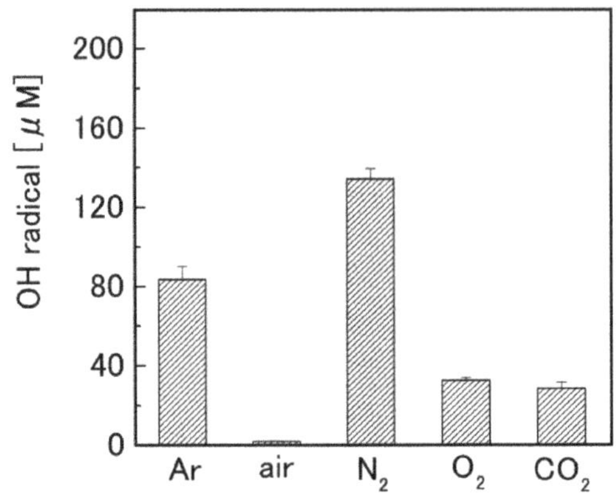

Figure 6.7 Amount of OH radicals produced by different gas plasmas (DMPO = 200 mM; plasma treatment time = 30 s).

Ozone measurement of various gas plasmas photometric measurement

Some reactive species form coloured reaction products, which can be identified by photometric measurements. The amount of generated ozone was measured by an ultraviolet/visible (UV–Vis) spectrometer (U-2010; Hitachi High-Technologies Co.). The ozone concentration in the liquid was measured by an ozone test kit (Hach Company). The reaction between the reagent and ozone in the solution yields a characteristic absorbance at 310 nm. By this method, ozone generated by plasmas of argon, mock air, nitrogen, oxygen, and carbon dioxide was investigated. Sample solutions of 200 μl were placed with their liquid surfaces 6 mm distant from the plasma source outlet and subjected to 15 s plasma treatment.

As shown in Fig. 6.8, the largest amount of ozone of 64 μM was generated by oxygen plasma. The ozone generated by plasmas of argon, mock air, nitrogen, and carbon dioxide plasma were 33, 9.9, 33, and 48 μM, respectively.

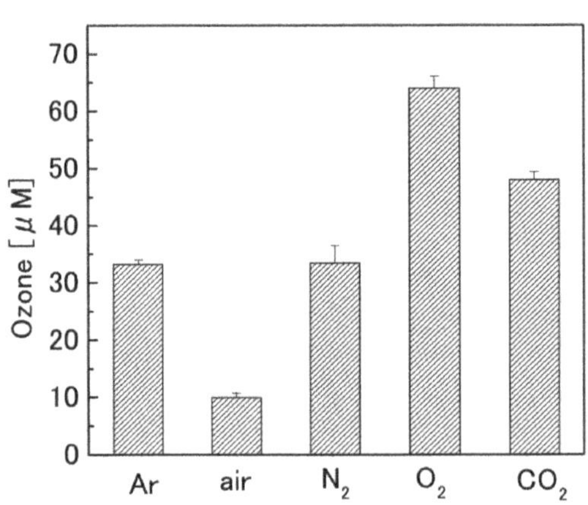

Figure 6.8 Amount of ozone produced by different gas plasmas (plasma treatment time = 15 s).

Discussion

The atomic, ion, and molecule lines of each gas species in the plasmas has already been confirmed (Takamatsu *et al.*, 2013). These reactive gas species react with each other and with water to produce reactive oxygen species (ROS) such as OH radicals, ozone, and other reactive species. ROS is thought to be capable of oxidizing organics to carbon dioxide and water (Kuroki *et al.*, 2006). In this study, treatment with nitrogen and oxygen, which generate large amounts of OH radicals and ozone molecules, efficiently reduced the intensity of the TTX peaks.

Regarding reactive-species production by nitrogen plasma, it appears that atomic nitrogen and water generate OH radicals as shown in equation 6.1 (Laroussi and Leipold, 2004). Nitrogen generated more OH radicals than the other gas plasmas, suggesting that OH radicals are mainly responsible for TTX degradation.

$$2N + 2H_2O \rightarrow N_2 + 2HO\cdot + 2H\cdot \tag{6.1}$$

Regarding reactive species production by oxygen plasma, it appears that atomic oxygen, water and molecular oxygen generate OH radicals and ozone as shown in equations 6.2 and 6.3, respectively (Takamatsu *et al.*, 2012; Ionin *et al.*, 2007), suggesting that ozone also decomposes TTX.

$$O + H_2O \rightarrow 2HO\cdot \tag{6.2}$$

$$O + O_2 + M \rightarrow O_3 + M \tag{6.3}$$

In contrast, treatment with mock air plasma, which produces low amounts of OH radicals and ozone, scarcely altered the TTX intensity. Although mock air contains both oxygen and nitrogen, these gases quickly react to yield NO radicals rather than OH radicals or ozone molecules. The reaction is shown in equation 6.4 (Herron and Green, 2001).

$$N + O \rightarrow NO \tag{6.4}$$

Future trends

From these findings, we expect that plasma treatment can neutralize organically bonded toxic substances in foods and drink, provided that the plasma gas species are suitable for decomposition.

References

Ionin, A.A., Kochetov, I.V., Napartovich, A.P., and Yuryshev, N.N. (2007). Physics and engineering of singlet delta oxygen production in low-temperature plasma. J. Phys. D: Appl. Phys. *40*, R25–R61.

Isbary, G., Heinlin, J., Shimizu, T., Zimmermann, J.L., Morfill, G., Schmidt, H.-U., Monetti, R., Steffes, B., Bunk, W., Li, Y., Klaempfl, T., Karrer, S., Landthaler, M., and Stolz, W. (2012). Successful and safe use of 2 min cold atmospheric argon plasma in chronic wounds: results of a randomized controlled trial. Br. J. Dermatol. *167*, 404–410.

Herron, J.T., and Green, D.S. (2001). chemical kinetics database and predictive schemes for nonthermal humid air plasma chemistry. Part II. Neutral Species Reactions: Plasma Chem. Plasma Process. *21*, 459–481.

Kohno, M., Yamada, M., Mitsuta, K., Mizuta, Y., and Yoshikawa, T. (1991). Spin-trapping studies on the reaction of iron complexes with peroxides and the effects of water-soluble antioxidants. Bull. Chem. Soc. Jpn. *64*, 1447.

Kumagai, W., Miyahara, H., Hotta, E., and Okino, A. (2007). Atmospheric line remote plasma source for surface treatment. The 6th Asian-European Int. Conf. on Plasma Surface Eng. *310*, P4007.

Kuroki, T., Yoshida, K., Watanabe, H., Okubo, M., and Yamamoto, T. (2006). Decomposition of trace phenol in solution using gas–liquid interface discharge. Japan. J. Appl. Phys. *45*, 4296–4300.

Laroussi, M., and Leipold, F. (2004). Evaluation of the roles of reactive species, heat, and UV radiation in the inactivation of bacterial cells by air plasmas at atmospheric pressure. Int. J. Mass Spectrom. *233*, 81–86.

Nakashima, N., Shibata, M., Takamatsu, T., Sasaki, R., Miyahara, H., and Okino, A. (2012). High-speed reduction of oxide film and surface hydrophilization using atmospheric multi-gas plasma jet. ESCAMPIG XXI, 4.

Ragni, L., Berardinelli, A., Vannini, L., Montanari, C., Sirri, F., Guerzoni, M.E., and Guarnieri, A. (2010). Non-thermal atmospheric gas plasma device for surface decontamination of shell eggs. J. Food Eng. *100*, 125–132.

Shigeta, K., Köllensperger, G., Rampler, E., Traub, H., Rottmann, L., Panne, U., Okino, A., and Jakubowski, N. (2013). Sample introduction of single selenized yeast cells (*Saccharomyces cerevisiae*) by micro droplet generation into ICP-sector field mass spectrometer for label free detection of trace elements. J. Anal. At. Spectrom. *28*, 637–645.

Shimizu, T., Steffes, B., Pompl, R., Jamitzky, F., Bunk, W., Ramrath, K., Georgi, M., Stolz, W., Schmidt, H.U., Urayama, T., Fujii, S., and Morfill, G.E. (2008). characterization of microwave plasma torch for decontamination. Plasma Process. Polym. *5*, 577–582.

Takamatsu, T., Ichikawa, M., Hirai, H., Sasaki, R., Shibata, M., Miyahara, H., Matsumoto, Y., and Okino, A. (2011). Sterilizing effect of various gases by damage-free multi-gas plasma jet. The 38th IEEE Int. Conf. on Plasma Sci., IP2H-23.

Takamatsu, T., Kawate, A., Uehara, K., Oshita, T., Miyahara, H., Dobrynin, D., Fridman, G., Fridman, A., and Okino, A. (2012). Bacterial inactivation in liquids using multi-gas plasmas. Plasma Medicine *2*, 237–247.

Takamatsu, T., Hirai, H., Sasaki, R., Miyahara, H., and Okino, A. (2013). surface hydrophilization of polyimide films using atmospheric damage-free multigas plasma jet source. IEEE Transactions on Plasma Sci. *41*, 119–125.

Tamura, T., Kaburaki, Y., Sasaki, R., Miyahara, H., and Okino, A. (2011). Direct decomposition of anesthetic gas exhaust using atmospheric pressure multigas inductively coupled plasma. IEEE Transactions on Plasma Sci. *39*, 1684–1688.

Tsai, Y.H., Hwang, D.F., Cheng, C.A., Hwang, C.C., and Deng, J.F. (2006). Determination of tetrodotoxin in human urine and blood using C18 cartridge column, ultrafiltration and LC–MS. J. Chromatogr. B. *832*, 75–80.

Watanabe, T., and Tsuru, T. (2008). Water plasma generation under atmospheric pressure for HFC destruction, Thin Solid Films *516*, 4391–4396.

Xie, J., Xin, D., Cao, H., Wang, C., Zhao, Y., Yao, L., Ji, F., and Qiu, Y. (2011). Improving carbon fibre adhesion to polyimide with atmospheric pressure plasma treatment. Surf. & Coat. Technol. *206*, 191–201.

Current Progress in Advanced Research into Fungal and Mycotoxin Inactivation by Cold Plasma Sterilization

7

Pervin Başaran Akocak

Abstract

Growth of fungi can cause physico-chemical spoilage, deterioration, nutritional and organoleptic property losses in food and feed commodities. Furthermore, mycotoxins produced by fungi are tremendous food safety and economic concern, and the maximum contamination levels are regulated by the international organizations. Mycotoxin control strategies include pre- and post-harvest detoxification approaches. Currently available methods are not sufficient for the full elimination or decontamination of mycotoxins. Mild non-thermal disinfection methods, which aim at actively improve the storability of goods and preserve the quality parameters of the product during post-harvest storage are continuously sought by the food industry. Recent studies indicate that application of plasma is among the promising, gentle non-thermal technologies and a new tool for the prevention and decontamination of fungal development and mycotoxin contamination while improving storability of food commodities during post-harvest storage.

Introduction

Food and feed commodities may become contaminated by filamentous fungi while in the field, during harvest handling, post-harvest storage and processing. Fungi cause nutritional loss, colour change, unpleasant odors, reduced digestibility and germination quality of seeds, baking and malting quality loss, and harm plant material's use in animal feed or food chain (FAO, 1996). Even further, some of the fungal species produce poisons named, mycotoxins which are tremendously affecting the safety, commercial value and final usability of the product. Nearly 4.5 billion people in developing countries are chronically exposed to critical amounts of mycotoxins (Williams *et al.*, 2004). Major factors that affect the mycotoxin production are plant genotypes, fungal species and the spore load, use of chemical preservatives, plant composion, soil type, unbalanced fertilization, insect activity, drought, humidity, temperature, oxygen level and atmospheric conditions during drying and storage (Magan and Aldred, 2007). There is no biochemical significance for mycotoxin production in fungal development and metabolism; and little is known about the complex factors (e.g. agronomic practices and climatic conditions) that influence the genetic regulation of mycotoxin biosynthesis (Yu and Keller, 2005).

Mycotoxins are chemically diverse secondary fungal metabolites, and generally have aromatic lactone or coumarin rings in their chemical structure. To date more than 300 chemically diverse mycotoxins produced by more than 100 moulds have been identified. However, for foodstuff and feedstuffs, only a small number of mycotoxins (aflatoxins, ochratoxins, trichothecenes (deoxynivalenol, nivalenol), zearalenone, fumonisins, ochratoxins and fumonisins) are of interest, and they are mainly produced by filamentous fungi in the genera of *Aspergillus* spp., *Penicillium* spp., *Fusarium* spp., *Claviceps* spp. and *Alternaria* spp. (Kumar *et al.*, 2007) (Table 7.1). Aflatoxins B1, B2, G1 and G2 are four naturally occurring forms of aflatoxins, mostly produced by *A. flavus, A. parasiticus, A. nominus* and *A. pseudotamari* (Basaran *et al.*, 2008). Aflatoxin B1 is excreted in milk in the form of aflatoxin M1 (Basaran *et al.*, 2008). Zearalenone is oestrogenic mycotoxin, biosynthesized through a polyketide pathway by a variety of *Fusarium* species (*F. graminearum, F. culmorum, F. cerealis, F. equiseti, F. crookwellense* and *F. semitectum*) (Pfohl-Leszkowicz *et al.*, 1995). Trichothecenes

Table 7.1 Major mycotoxins of food and feed concern and the food produces in which they are reported

Mycotoxins	Producing fungal species	Food/feed produces
Aflatoxins	*Aspergillus flavus, A. parasiticus, A. nominus, A. pseudotamari*	Tree nuts, fig, peanut, spice pepper, other spices, cereals (maize, barley), beer, cotton seed, milk (animal consumption of mould/toxin contaminated feed)
Ochratoxin	*Aspergillus ochraceus, A. carbonartus, A. niger, A. alliaceus, A. sclerotiorum, A. sulphureus, A. albertensis, A. auricomus, A. wentii, Penicillium verrucosum*	Grape, raisins, cacao, nut, spices, cereals, wine, coffee, spices, beans, groundnuts, milk and meat products (animal consumption of mould/toxin-contaminated feed), beer
Zearalenone	*Fusarium culmorum, F. graminearum, F. graminearum, F. culmorum, F. cerealis, F. equiseti, F. crookwellense, F. semitectum*	Maize, barley, oats, wheat, sorghum, millet, rice, cereals, soya, beer, acha, dried fruits and vegetables
Fumonisins	*Fusarium vericillioides, F. proliferatum*	Maize, sorghum, cereals, flour, starch, groat, onion, garlic, asparagus, pea seed
Trichothecenes	*Fusarium culmorum, F. oxysporium, F. solani, F. equiseti, F. graminearum, F. moniliforme, F. pseudograminearum, F. sporotrichioides, Trichotceum roseum, F. poae, F. cerealis*	Grains, fruits, dried fruit products, beans, wheat, barley, maize, cereals, rye, soybean
Patulin	*Penicillium expansum, P. patulum, Aspergillus clavatus, A. giganteus, Byssochlamys nivea*	Meat and eggs (animal consumption of mould/toxin contaminated feed), fruit juice, fruit (apple, pear, peach, cherry, apricot, pineapple, grape, banana, strawberry, plum) and fruit derived products
Fusarins	*Fusarium acuminatum, F. armentacum, F. culmorum, F. graminearum, F. culmorum*	Maize, cereals, wheat, barley, oat
Culmorin	*Fusarium crookwellense, F. sporotrichioides, F. graminearum, F. culmorum*	Cereals, rye, sorghum, malt

are sesquiterpenoids, mainly produced by species of *Fusarium* and fungal genera like *Trichoderma*. Their biosynthesis begins with the formation of trichodiene, which undergoes oxygenation and esterification reactions (Ward *et al.*, 2002). Deoxynivalenol belongs to the group of trichothecenes and mostly found in cereal grains. Patulin is produced by particularly *Penicillium* spp., *Aspergillus* spp. and *Byssochlamys* spp. species, which commonly occur in fruits and cereals (Steiman *et al.*, 1989). Fumonisin was first isolated from South Africa in 1988 and formed mainly by *Fusarium verticillioides* and *Fusarium proliferatum* (Yoshizawa, *et al.*, 1994). *Fusarium* spp. produced zearalenone contaminates maize, barley, oat, wheat, sorghum, millet and rice, and processed cereals products such as malt, soy souce and beer (CCFAC, 2000). Varga *et al.* (1996) reported detailed screening of ochratoxins forming species, which are commonly detected in dried fruits, cereals, milk, meat products, and spices.

Mycotoxins evoke toxic response and significant hazard when introduced even in very low concentration to wide range of animals and humans by ingestion, touching and absorbsion through the skin contact, and inhalation through lungs (Zain, 2011). Mycotoxic effects called mycotoxicoses might be acute, subacute or long-term chronic. Aflatoxins are categorized as Group 1 carcinogens (IARC, 2002). They inhibit transcription and protein synthesis, and even further they cause mutations and teratogenic effect, neural tube defects in newborn infants, anorexia, lethargy, jaundice, esophageal cancer, can impair the central nervous system, and compromise immunological system (Sharma, 1993; Williams *et al.*, 2004; Wild and Turner, 2002; Mishra and Das, 2003). Once digested, aflatoxins are converted by the liver into toxic reactive epoxides which bind covalently to macromolecules such as DNA, RNA and enzymes, resulting in hepatic damage to liver cells, enlarged livers, disseminated intravascular coagulation and internal haemorrhaging (Williams *et al.*, 2004; Cullen and Newberne, 1994; Pereyra *et al.*, 2008). Furthermore, in animals, aflatoxin B1 is associated with several diseases, such as leukoencephalomalacia in mammalians, skeletal anomalies in rabbits, vomiting, depression, polydipsia, polyuria, anorexia, weakness and diarrhoea in cats and dogs, and liver cancer and renal necrosis in rats (Newberne *et al.*, 1966; Schmidt and Panciera; 1980; Miller and Wilson, 1994). Patulin toxicity to mammals includes genotoxicity, teratogenicity, mutagenicity, embryotoxicity and carcinogenicity (Donmez-Altuntas *et al.*, 2013; Roll *et al.*, 1990). Zearalenone is an oestrogenic resorcylic acid lactone that causes severe morphological and functional disorders of reproductive organs in livestock; additionally, hepatocellular adenomas and pituitary tumours have been observed in long-term carcinogenicity studies in mice (Richard, 2007; Maragos *et al.*, 2010). Ochratoxin demonstrates nephrotoxic, teratogenic and immune-suppressive properties. It causes cancer and deterioration of liver or kidney function (O'Brien *et al.*, 2005). Fumonisin has been demonstrated to induce apoptosis in cultured human cells and in rat kidneys, and triggers human oesophageal carcinoma, and furthermore its consumption has been implicated in neural tube defects in babies (Yoshizawa *et al.*, 1994; Tollenson *et al.*, 1996; Marasas *et al.*, 2001). Deoxynivalenol reduces feed intake and suppresses the immune system in farm animals, activates critical cellular kinases involved in signal transduction, and inhibits protein synthesis; further, it causes circulatory shock, reduced cardiac output and, ultimately, death (Doll *et al.*, 2008; Pestka and Smolinski, 2005). Because of above-mentioned effects stringent regulations and maximum tolerated limits were imposed worldwide. Most instantly, the European Food Safety Authority (EFSA) guidance values are given with 0.5 mg zearalenone/kg and 5 mg deoxynivalenol/kg relative to feed with 12% moisture content (Winkler *et al.*, 2013).

Conventional inactivation methods for fungi and mycotoxins

The occurrence of fungal development and accumulation of mycotoxins in food and feedstuff are not avoidable. Regardless of all the precautions, the Food and Agricultural Organization (FAO) of the United Nations estimates nearly a quarter of the world's plant crops are contaminated with mycotoxins to various extents (WHO, 1999). Management of mycotoxin contamination includes prevention of exposure to mycotoxin-producing fungi, monitoring of mould development, routinely analysis of mycotoxins, and finally sustainable decontamination. There are a number of prevention strategies (physical, chemical and biological methods or combinations of these) for contamination and the growth of fungi in the field and post-harvest. Pre-harvest strategies include crop rotation, sowing date, tillage, cultivation techniques, selection of soil fertilizers, crop variety selection and breeding, use of transgenic species resistance to development, pest control, fungicide applications, and microbial strains (e.g. endophytic bacteria or atoxigenic fungi) that can out-compete toxigenic strains in the field (Bacon *et al.*, 2001; Jouany, 2007). According to Dorner *et al.* (2002), atoxigenic strains of *A. flavus* and *A. parasiticus* introduced to soil of developing peanut crops reduced aflatoxin contamination in the following year by 70%. Ochratoxin and patulin producing fungi were reported to be inhibited by the yeast species of *Candida zemplinina, Saccharomyces cerevisiae, Pichia kluyveri,* and *Metschnikowia aff. fructicola* (Zhu *et al.*, 2015a,b).

Despite all the pre-harvest efforts to prevent fungal contamination in the field, mould spores can still contaminate and then grow during storage and transportation of agricultural products (Basaran *et al.*, 2008). Successful tools are sought for the post-harvest control of pathogenic fungi. Post-harvest control schemes consist of effective drying, control of humidity and temperature during storage, physical removal of contaminated materials by sorting, washing, dehulling, use of chemicals and preservatives, and storage under modified atmosphere conditions. Fumigation with ethylene oxide, prochloraz, propiconazole, epoxyconazole, tebuconazole, cyproconazole, itraconazole, amphotericin B, and azoxystrobin can inhibit fungal metabolism and fungal development (Haidukowski *et al.*, 2004). However, chemical contamination is not desired by the consumers and strictly regulated by national food safety authorities, and the fumigation with some compounds are prohibited in many countries due to detrimental effects to human health and environment (Fowles *et al.*, 2001). The effectiveness of chlorine water surface spraying is limited, and there is the risk of formation of carcinogenic chlorinated compounds, and therefore its usage on minimally processed vegetables is restricted in the Netherlands, Sweden, Germany and Belgium (Rico *et al.*, 2007). Several studies have also reported fungicidal effects of fumigation with plant essential oils extracted from herbs or plants such as cinnamon, citral, palmarose, eugenol oil, *Litsea cubeba*, clove, eucalyptus, anise, spearmint and camphor, but their use is not practically applicable (Chao *et al.*, 2000; Velluti *et al.*, 2003). Although 2–7 kGy gamma irradiation appears to be partially effective to decontaminate fungi, and its implementation has been strongly obstructed worldwide. Fungi are known to be highly resistant to irradiation, it causes nutritional losses in food, consumers have negative opinion of irradiation, and there is strict international legal regulation for the irradiated ingredients (Arvanitoyannis *et al.*, 2010). UV-C irradiation has been applied as a non-thermal treatment to eliminate mycotoxin-producing fungi without adversely changing the quality of nuts (Basaran, 2009). But UV-C showed limited penetration on the surface, fungal inactivation was not homogeneous, and the ineffectiveness of the UV-C radiation is dependent on the power and the distance from the UV source (Basaran, 2009).

Regardless of fungal decontamination strategies; mycotoxins frequently contaminate agricultural commodities, so successful detoxification tools are still researched to reduce or eliminate the toxic effects of mycotoxins. Food related mycotoxins are extremely resistant to high temperature, food processes such as baking, roasting, frying, extrusion shown to have only partial inhibition effect (Bullerman and Bianchini, 2007). Roasting pistachio nuts at 150°C for extreme 120 min could have reduced the aflatoxin content only 63%, while pressure cooking of rice contaminated with aflatoxin B1 showed a reduction of 70% (Park *et al.*, 2005; Yazdanpanah *et al.*, 2005). Cenkowski *et al.* (2007) studied the effect of superheated steam at 185°C and observed 50% reduction in deoxynivalenol after 6 min. Baking of biscuits could have destroyed only about two-thirds of the ochratoxin (Subirade, 1996). Although some promising results were observed with heat processing, high temperature causes unacceptable changes in nutritional and sensory properties of the product. There are contradicting reports on the impact of irradiation on mycotoxin levels. Irradiation of a 50-mg aflatoxin solution with 10 kGy resulted in 40% inactivation (Patel *et al.*, 1989). Markov *et al.* (2015) observed that 5 kGy irradiation reduced aflatoxin B1 by around 60% of initial contamination, while some vitamins such as tiamin and niacin are not stable when treated with ionizing radiation (Khattak and Klopfenstein, 1989). Ritieni *et al.* (1999) and Basaran (2009) studies UV irradiation effect on mycotoxins. Ritieni *et al.* (1999) reported no effect on fusaproliferin, produced by some phytopathogenic *Fusarium* species; while Basaran *et al.* (2009) observed that 6 h UV-C yielded nearly 25% reduction in aflatoxin B1 and G1, but this dose had no effect on aflatoxin B2 and G2 on contaminated hazelnuts. Mycotoxins can be partially destroyed or converted to less toxic compounds by chemical treatment. These chemicals include acids, bases, hydrogen peroxide, ozone and other oxidizing agents, citric acids, bisulphites, $NaHCO_3$, H_2O_2, calcium hydroxide monoethylamine, ammonia or calcium hydroxi demonoethylamine (Applebaum and Marth, 1982; Coker *et al.*, 1998; Park, 1993; Bauer, 1994; Moerck *et al.*, 1980; Mendez-Albores, 2007; Gwenaelle *et al.*, 2011). Beekrum *et al.* (2003) reported that naturally occurring phenols (chlorophorin, irokom, maakianin, vanillic acid, and caffeic acid) were effective in the degradation of aflatoxin B1. Efficiency of the chemical degradation is specific to the structure of mycotoxin under investigation. Concentrated ozone has completely degraded and detoxified zearelone, but a greater resistance of aflatoxin B2 and G2 was observed (McKenzie *et al.*, 1997). The presence of double bonds at C8–C9 position for aflatoxin B1 and aflatoxin G1, and the tendency of ozone to react at olefinic positions indicated the possible sensitivity towards ozone (Cullen *et al.*, 2009). The most recent approach reducing mycotoxins in the feed industry is the addition of nutritionally inert sorbent materials and removal of mycotoxins by adsorption or binding to polymers (Avantaggiato *et al.*, 2005). Aluminosilicates (e.g. clays, zeolites) were reported for the adsorption of mycotoxins into porous structure and trapping by electric elementary charges (Colvin *et al.*, 1989). The limitation of aluminosilicate adsorption is the simultaneous decrease of quality parameters such as colour, aroma, and flavour characteristics. Biotransformation, biodegradation or fermentation of the mycotoxins by a variety fungi (*A. niger, Trichoderma viride, Mucor ambiguus, Eurotium herbariorum, Rhizopus* spp.), yeast (*Trichosporon mycotoxinivorans, Xanthophyllomyces dendrorhous, Rhodosporidium paludigenum, Pichia anomala, P. kluyveri and Hanseniaspora uvarum*), and bacteria (*Mycobacterium fluoranthenivorans, Rhodococcus* spp., *Corynebacterium rubrum, Lactobacillus rhamnosus, Rhodococcus erythropolis, Flavobacterium auranticum*) to less toxic compounds have been reported (European Food Safety Authority, 2009; Zhu *et al.*, 2015; El-Nezami *et al.*, 2000; Hormisch *et al.*, 2004; Masoud and Kaltoft, 2006; Yin *et al.*, 2008). According to Karlovsky (1999), biological control or bioconversion of mycotoxins to non-toxic substances is usually incomplete or inefficient.

Non-thermal (cold) plasma applications

Mycotoxins are small molecules that are extremely difficult to remove or eliminate without harming the produce. Currently available methods are not fully sufficient to combat mycotoxin-producing fungi, asthey have significantdrawbacks. Some methods cause unavoidable chemical changes, have unacceptable negative effects on the final products and do not completely eliminate fungal growth and mycotoxin contamination, and some of these processes are not economically viable. So far, none of the above-mentioned strategies has been able to reduce mycotoxin production to acceptable levels. Gentle and more effective techniques are considered necessary by food handlers to deal with fungal contaminants.

Plasma refers to neutral ionized (or energized) gases, composed of photons, ions and free electrons as well as atoms in their fundamental or excited states with a net neutral charge (Selcuk *et al.*, 2008). Cold plasma is generated by excitation of gases with an immense energy (high-voltage electrical discharges, microwaves, irradiation or other energy sources) under either low or atmospheric pressure. In the literature, there are different approaches for the design of a plasma production/treatment unit for food products. But in general, gas, power source, plasma production and treatment unit, and vacuum system are included in a food related plasma treatment system. Fig. 7.1 shows one of the examples for a plasma treatment unit which has been used in our earlier studies (Basaran *et al.*, 2008).

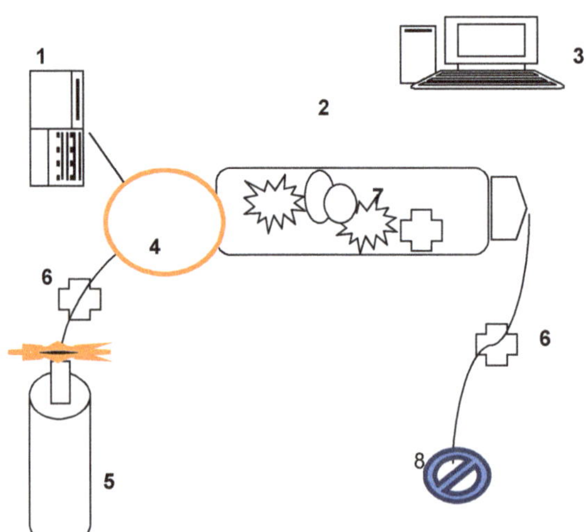

Figure 7.1 Plasma sterilization unit. 1, power uni; 2, treatment/sterilization compartment; 3, Remote computer control system; 4, plasma production compartment; 5, gas source; 6, sensors; 7, food; 8, vacum pump.

Gas plasma is efficient to inactivate a wide range of microorganisms (bacterial vegetative cells, spores, yeasts and moulds) by adhering to foodstuff surface within short treatment durations and at ambient temperatures (Akitsu *et al.*, 2005). In order to produce plasma effective for the elimination of microorganisms, conditions are to be optimized. The major factors that affect the competence are plasma source gas species, treatment duration, electric pulse and power, microbial species, and surface properties of food item. Antimicrobial effect of plasma sterilization has been reported with a number of gases such as O_2, N_2, air, H_2, N_2O_3, H_2O_2, CO_2, SO_2, halogens, and SF_6, etc. (Kim *et al.*, 2014, Selcuk *et al.*, 2008; Scholtz *et al.*, 2015; Ehlbeck *et al.*, 2011; Shintani *et al.*, 2010). The definite effect mechanism of plasma is complex and varies substantially with the type and chemical composition of reactive species. If air is used as gas source, a mixture of neutral and reactive oxygen species (O_2^+, O_2^-, O_3, O, O^+, O^-), charged nitrogen species such as N^+, NO^+, NO^-, NO_2, N_2O_4, N_2O), hydroxyl and hydroperoxyl radicals (OH and HO_2), hydrogen peroxide (H_2O_2) are produced (Scholtz *et al.*, 2015; Kim *et al.*, 2014). Oxygen species generally demonstrate germicidal effect by etching action on cell membrane of food pathogens. Surface amino acids of glycoproteins and lipids are oxidized by electrons and ions, which leads to diffusion of cellular contents, growth inhibition, and ultimate death of the cells (Zhang and Chen, 2009; Kvam *et al.*, 2012). Ohkawa *et al.* (2006) observed that the spherical cellular structure of yeast *C. albicans* was destroyed, and the residue of cell leakage was detected after cold atmospheric plasma treatment. In other studies, cracking of cell walls and broken conidiophores and the vesicle of fungal species were observed (Basaran *et al.*, 2009; Yang *et al.*, 2009; Suhem *et al.*, 2013). Even further some studies indicate damage to DNA chain, metabolic disturbance in biosynthesis, reproduction and repair mechanisms (Akishev *et al.*, 2010).

Although there have been quite a number of studies reporting antibacterial effect of plasma, limited number of studies focused on the effect of plasma against fungi and mycotoxins. Studies to date provide evidence that while no significant differences in susceptibility among bacterial species is observed under same treatment conditions, fungi requires longer treatment duration (20–30 min) and usually antifungal effect significantly depends on the plasma system, the type of gas used for plasma production, fungal species, microbial load, and the food's composition and the surface condition (Basaran *et al.*, 2008; Scholtz *et al.*, 2015). *A. niger* spores were spread on an agar medium surface and exposed to the atmospheric pressure plasma with variable nitrogen and oxygen mixtures. Inhibition zones of 30–40 mm in diameter after 30–60 s of treatment were observed (Akishev *et al.*, 2008). In our earlier studies, antifungal effects of air plasma and SF_6 gas plasmas on hazelnut, peanut and pistachio nuts were evaluated in detail (Basaran *et al.*, 2008). Air plasma treatment at 10 min resulted in 2 log reduction of *A. parasiticus*. Determination of the mould viable counts from SF_6 treated hazelnuts demonstrated nearly a 5 log decrease in CFU after 5 min of treatment, any further duration of SF_6 plasma treatment did not significantly affect the fungal load (Basaran *et al.*, 2008). Suhem *et al.* (2013) reported the effects of argon plasma treatment on the growth of *A. flavus* on malt extract agar and brown rice cereals at powers of 20 W and 40 W with exposure times at 5, 15 and 25 min. No germination of *A. flavus* was observed after a plasma treatment at 40 W for 25 min. The structure of *A. flavus* was damaged by argon plasma where conidiophores and the vesicle were found to be broken (Suhem *et al.*, 2013). Effects of the microwave-powered cold plasma treatments on the inhibition of *A. flavus* artificially inoculated red pepper powder were reported by Kim *et al.* (2014). *A. flavus* was reduced by 2.5 log spores/gram with nitrogen plasma at 900 Watt after 20 min duration

(Kim *et al.*, 2014). They concluded that the reactive nitrogen species and UV photons, electrons, and hydroxyl radicals might have oxidized glycoproteins in the fungal cell wall. The inactivation of *F. graminearum*, *F. oxysporum*, and *Neurospora crassa* using microwave plasma with nitrogen, argon and air gases were tested (Na *et al.*, 2013). Hyphal growth was most impeded in *F. graminearum* under same treatment duration as compared with other fungal species tested. Reduction in hyphal extension was much more when combination of Ar and O_2 was used, compared to combinations such as Ar and N_2, Ar and air, or N_2 (Na *et al.*, 2013). Park *et al.* (2012) reported the cellular and molecular responses of the filamentous *N. crassa* to the action of argon plasma, the plasma treatment did not show significant change for the germination rate in sodium chloride solutions, whereas more than 50% of *N. crassa* spores were inactivated in water or non-ionic sorbitol, glycerol and sucrose solutions (Kang *et al.*, 2014). *N. crassa* spores appeared to shrink and damage to cytoskeletal structures was observed. Park *et al.* (2012) and Kang *et al.* (2014) concluded that the surrounding environment enormously affect the behaviour of reactive species.

Seed treatment is the most cost-effective disease control method for agricultural commodities. Although a number of studies reported the plasma treatment for the sterilization of the seed surface; the experimental data on the seed germination and plant development after plasma treatment are scanty and disputed. Selcuk *et al.* (2008) determined the efficacy of a low pressure cold plasma system using air gases and SF_6 to reduce strains of *Aspergillus* spp. and *Penicillum* spp. on artificially inoculated seed surface and the treatment duration ranged from 5 to 20 min. A 3 log reduction of the initial load was observed for 15 min SF_6 plasma treatment time and damage to living seeds was avoided (Selcuk *et al.*, 2008). In other studies, root and sprout length and dry weight after germination were measured and concluded that non-thermal plasma treatment enhanced the germination and plant productivity of wheat, maize, soybean and other species within early growth time (Filatova *et al.*, 2012; Ling *et al.*, 2014). Even further, pre-sowing plasma treatment was reported to enhance germination and improve plant production (Selcuk *et al.*, 2008; Filatova *et al.*, 2012). Volin *et al.* (2000) conducted an earlier study where seeds were treated with tetra fluoride or octadecafluorodecalin. The final percentage of germination was almost the same in the cases of plasma-treated and untreated samples. However, they reported a delay in germination as compared with the untreated controls (Volin *et al.*, 2000).

Plasma treatment is also candidate to reduce natural microflora of yeast and mould contamination for the heat sensitive food packaging materials (Lee *et al.*, 2015). Muranyi *et al.* (2007) reported the microbial inactivation effectiveness of plasma against the fungus *A. niger* on polyethylene terephthalate foils. *A. niger* appeared to be the most resistant species with an inactivation rate of about 5 log in 5 s. Later, Muranyi *et al.* (2008) investigated the influence of humidity on the inactivation, and the *A. niger* spores were mostly inactivated at a high relative humidity of 70% (approx. 2 log). An alternative plasma source for treatment of foods is the use of a dielectric barrier discharge set-up, which offers the advantage of the treatment of food inside an already sealed container that prevents the risk of recontamination post processing (Misra *et al.*, 2014a,b). Matan *et al.* (2014) tested the combined effect of cold atmospheric plasma and clove oil on palm sheet as food packaging material, against growth of *A. niger*, *Penicillium* spp., and *Rhizopus* spp. Plasma increased the antifungal activity of clove oil; furthermore, wettability tests showed that plasma treatment increased the contact angle of the leaf sheath (Matan *et al.*, 2014). The results have demonstrated the good effect of using atmospheric plasma treatment to enhance antifungal activity of clove oil and to control moulds on the organic packaging material.

Heat-resistant mycotoxins do not decompose readily, and they are difficult to eliminate in foods without compromising sensorial and nutritional quality (Shapira and Paster, 2004). Elimination of fungi does not mean detoxification of mycotoxins, because some fungi are not able to produce mycotoxins and those producing mycotoxins secrete different amounts depending on the fungal species, the substrate and stress conditions (Basaran, 2009). Aflatoxin elimination effect of air and SF_6 plasmas was investigated (Basaran *et al.*, 2008). As compared with control samples the level of aflatoxin B1 and B2 was reduced nearly 88–90% with air plasma (Basaran *et al.*, 2008). Effect of SF_6 gas plasma was more moderate and aflatoxin B1 and B2 estimated to be reduced approximately 80–87%. Aflatoxin G1 and G2 were most resistant and only a maximum of 60% and 30% reduction were detected, repectively (Basaran *et al.*, 2008). Park *et al.* (2007) reported the degradation effect of microwave-induced argon plasma at atmospheric pressure on three mycotoxins (aflatoxin B1, deoxynivalenol and nivalenol). Deoxynivalenol and nivalenol were degraded relatively slowly as compared to aflatoxin B1, and a complete removal of the mycotoxins was observed after 5 s of plasma treatment (Park *et al.*, 2007). Most recently, Wang *et al.* (2015) investigated low pressure plasma's ability to inactivate aflatoxin B1, and 300 Watt plasma resulted in 88% degradation after 10 min of treatment. The possible structures of the degradation products were also elucidated by Wang *et al.* (2015). They proposed that the degradation of aflatoxin B1 was initiated with an addition reaction that occurred in the C8–C9 bond, resulting in the formation of an intermediate ($C_{17}H_{15}O_7$) as the major product. The inhibitory effect of cold argon plasma on fumonisin B2 and ochratoxin contaminated on date fruits were reported by Ouf *et al.* (2014). Fumonisin B2 was not detected after 6 min of plasma treatment, whereas ochratoxin was completely removed when the fungus was treated for 7.5 min.

While antifungal and anti mycotoxin effect of cold plasma are tested; the consequences of cold plasma treatment on the quality characteristics (general apperance, colour, firmness and texture, health properties, plant growth rate, plant metabolic activity, induction of plant secondary metabolites, nutritional yield, and photosynthetic operations) of fresh or processed fruit and vegetable products have also been assessed by various studies (Kim *et al.*, 2014; Lee *et al.*, 2004; Lacombe *et al.*, 2015; Grzegorzewski *et al.*, 2011; Misra *et al.*, 2014a,b; Baier *et al.*, 2013; Bubler *et al.*, 2015; Park *et al.*, 2013). Published studies have indicated some conflicting results. Quality of red pepper is attributed to the presence of red coloured capsanthin (Lee *et al.*, 2004). Plasma treatment did not significantly change the colour of pepper powder as compared with untreated samples (Kim *et al.*, 2014). In a later study, while 90 min plasma treatment had no significant effect on black pepper seed; significant colour changes due to the remote plasma treatment were observed for the red paprika powder and crushed oregano, the effect was independent of the treatment time (Hertwig *et al.*, 2015). They concluded that green colour change in oregano was due to destruction of chlorophyll. Grzegorzewski *et al.* (2011) studied the interactions of plasma reactive species with secondary plant metabolites in lettuce. They suggested that the combined interactions of argon and reactive oxygen species may lead to degradation of epidermal cells due to accumulation of flavonoids and other compounds in vacuoles. According to Baier *et al.* (2013), plasma exposure leads to a detrimental effect on tissue photosynthetic efficiency by erosion of upper epidermis in the leaves. Lacombe *et al.* (2015) treated berries with a mixture of cold air plasma. Treatments longer than 90 s resulted in significant reductions in firmness and colour of anthocyanins. Cold plasma produced bluer surface colour on the berry fruits, which indicated that treatment does not bleach the fruit despite the significant loss in anthocyanins after 90 s. Divergent results were obtained after argon plasma treatment on the content of anthocyanins. According to Matan *et al.* (2015) radio frequency induced argon plasma did not affect total phenolic content of fresh dragon fruit, while plasma treated sour cherry juice at optimized plasma conditions had higher anthocyanin (34%) and phenolic acid (15%) content as compared to pasteurized and untreated juice (Garofulic *et al.*, 2015). Misra *et al.* (2014a,b) evaluated the quality parameters of colour, firmness, pH and weight loss of cherry tomatoes and strawberries after dielectric barrier discharge air plasma and observed no adverse effects.

The most important enzymes involved in quality of fruits and vegetable products are lipoxygenase, hydrolylic enzymes, peroxidase, and polyphenoloxidase. Studies evaluating the efficacy of plasma on the enzymatic inactivation in fruits and vegetables are scarce. Surowsky *et al.* (2014) showed that argon and oxygen mixture plasma was capable of reducing the activity of both polyphenoloxidase and peroxidase enzymes, which are critical for the control of colour attributes of apple cider. The activity of polyphenoloxidase was reduced by about 90% after a treatment time of 180 s. Peroxidase was more stable and was reduced by about 85% after 240 s. Pankaj *et al.* (2013) demonstrated the applicability of in-package cold plasma technology as a means to inactivate of tomato peroxidase enzyme at 30, 40 and 50 kV, for up to 5 min of dielectric barrier discharge generated air plasma treatment, and with reduction in enzyme activity, improved quality of the fruits have been noted. Takai *et al.* (2012) studied the helium plasma and a mixture of nitrogen and oxygen plasma systems on egg white lysozyme. They reported a decrease in the lysozyme activity, possibly due to changes in the side chains of amino acids and secondary structure of the enzymes (Takai *et al.*, 2012).

The number of studies on the evaluation of organoleptic properties of plasma treated food products is limited. Basaran *et al.* (2008) investigated the organoleptic properties of nuts following surface treatment by SF_6 and air gasses. The results obtained from the sensory panels led to the conclusion that the quality attributes (odour, appearance, texture, and overall acceptance) of nuts treated with 20 min air gases or SF_6 plasma did not differ significantly from the untreated samples. When organoleptic properties of brown rice cereal were evaluated an odour like boiled rice was detected after an argon plasma exposure time of 25 min, and product was rejected by the consumer panelists (Suhem *et al.*, 2013).

Conclusions

Owing to the implications of mycotoxins for human health and farm ecology, the management of fungal invasion of agricultural commodities is a serious agronomic problem. There is crucial need to eliminate mycotoxin-producing fungi and mycotoxins in order to improve food/feed safety, prevent economic losses, and recover contaminated products as much as possible. Plasma presents a residue-free, harmless, eco-friendly, fast, and a superficially applicable approach to decontaminate fungal and mycotoxin contamination on thermally labile organic or inorganic materials.

Future trends

Human and animals are usually exposed to multiple mycotoxins in diet. It would be useful to develop decontamination strategies, which aim at more than one mycotoxin. At present the major concern of plasma application is the initial construction cost of equipment and the appliance cost for per treatment of plasma systems. More research efforts must be undertaken to evaluate affordable and sustainable batch or continuous plasma disinfection systems that does not harm nutritional, sensory and other quality aspects of edible materials. Food processing technologies are complex and more than one approach is usually accommodated for each processed food. The concept behind an integrated 'hurdle' effect is to minimize risk of each phase of food processing. Plasma application combined with other processing technologies may offer synergistic or additive effect to conventional methods for conservation of food while safeguarding the quality properties. The challenge now is to determine the effect of plasma combined technologies on the degradation of mycotoxins in real food systems.

References

Akishev, Y., Grushin, M., Karalnik, V., Trushkin, N., Kholodenko, V., Chugunov, V., Kobzev, E., Zhirkova, N., Irkhina, I., and Kireev, G. (2008). Atmospheric-pressure, nonthermal plasma sterilization of microorganisms in liquids and on surfaces. Pure Appl. Chem. *80*, 1953–1969.

Akitsu, T., Ohkawa, H., Tsuji, M., Kimura, H., and Kogama, M. (2005). Plasma sterilization using glow discharge at atmospheric pressure. Surf. Coat. Technol. *193*, 29–34.

Applebaum, R.S., and Marth, E.H. (1982). Inactivation of M1 in milk using hydrogen peroxide and hydrogen peroxide plus riboflavin of lacto peroxide. J. Food Protect. *45*, 557–560.

Arvanitoyannis, I.S., Stratakos, A.C., and Tsarouhas, P. (2009). Irradiation applications in vegetables and fruits: a review. Crit. Rev. Food Sci. Nutr. *49*, 427–462.

Avantaggiato, G., Solfrizzo, M., and Visconti, A. (2005). Recent advances on the use of adsorbent materials for detoxification of Fusarium mycotoxins. Food Add. Contam. *22*, 379–388.

Bacon, C.W., Yates, I.E., Hinton, D.M., and Meredith, F. (2001). Biological control of *Fusarium moniliforme* in maize. Environ. Health Perspect. *109*, 325–332.

Baier, M., Foerster, J., Schabel, U., Knorr, D., Herppich, W.B., and Schlüter, O. (2013). Direct non-thermal plasma treatment for the sanitation of fresh corn salad leaves: evaluation of physical and physiological effects and antimicrobial efficacy. Postharvest Biol. Technol. *84*, 81–87.

Basaran, P., Akgul, N.B., and Oksuz, L. (2008). Elimination of *Aspergillus parasiticus* from nut surface with low pressure cold plasma (LPCP) treatment. Food Microbiol. *25*, 626–632.

Basaran, P. (2009). Reduction of *Aspergillus parasiticus* on hazelnut surface by UV-C treatment. Int. J. Food Sci. Technol., *44*, 1857–1863.

Bauer, J. (1994). Möglichkeiten zur Entgiftung mykotoxinhaltiger Futtermittel. Monatshefte für Veterinärmedizin. *49*, 175–181.

Beekrum, S., Govinden, R., Padayachee, T., and Odhav, B. (2003). Naturally occurring phenols: A detoxification strategy for fumonisin B1. Food Additives and Contaminants *20*, 490–493.

Bubler, S., Herppich, W.B., Neugart, S., Schreiner, M., Ehlbeck, J., Rohn, S., and Schlüter, O. (2015). Impact of cold atmospheric pressure plasma on physiology and flavonol glycoside profile of peas (*Pisum sativum* 'Salamanca'). Food Res. Int. doi:10.1016/j.foodres.2015.03.045

Bullerman, L.B., and Bianchinii, A. (2007). Stability of mycotoxins during food processing. Int. J. Food Microbiol. *119*, 140–146.

CCFAC. (2000). Committee on Food Additives and Contaminants. Posting date. Joint FAO/WHO Expert Committee on Food Additives: Position paper on zearalenone. Publication CCFAC 00/19. Codex Alimentarius Commission, Rome, Italy, 2000.

Cenkowski, S., Pronyk, C., Zmidzinska, D., and Muir, W.E. (2007). Decontamination of food products with superheated steam. J. Food Eng. *83*, 68–75.

Chao, S.C., Young, D.G., and Oberg, C.J. (2000). Screening for inhibitory activity of essential oils on selected bacteria, fungi and viruses. J. Essential Oil Res. *12*, 630–649.

Coker, R.D. (1998). The chemical detoxification of aflatoxincontaminated animal feed. Nat. Toxicants Food, 284–298. Sheffield Academic Press, Boca Raton, FL.

Colvin, B.M., Sangster, L.T., Haydon, K.D., Beaver, R.W., and Wilson, D.M. (1989). Effect of a high affinity aluminosilicate sorbent on prevention of aflatoxicosis in growing pigs. Vet. Human Toxicol. *31*, 46–48.

Cullen, P.J., Tiwari, B.K., O'Donnell, C.P., and Muthukumarappan, K. (2009). Modelling approaches to ozone processing of liquid foods. Trends Food Sci. Technol. *20*, 125–136.

Cullen, J.M., and Newberne, P.M. (1994). Acute hepatotoxicity of aflatoxins, In: The toxicology of aflatoxins: human health, veterinary and agricultural significance. Eaton, D.L. & Groopman, J.D., pp. 3–26, Academic Press, San Diego, USA.

Doll, S., Danicke, S., and Valenta, H. (2008). Residues of deoxynivalenol (DON) in pig tissue after feeding mash or pellet diets containing low concentrations. Mol. Nutr. Food Res. *52*, 727–734.

Donmez-Altuntas, H., Gokalp-Yildiz, P., Bitgen, N., and Hamurcu, Z. (2013). Evaluation of genotoxicity, cytotoxicity and cytostasis in human lymphocytes exposed to patulin by using the cytokinesis-block micronucleus cytome (CBMN cyt) assay. Mycotoxin Res. *29*, 63–70.

Dorner, J.W., and Cole, R.J. (2002). Effect of application of nontoxigenic strains of *Aspergillus flavus* and *A. parasiticus* on subsequent aflatoxin contamination of peanuts in storage. J, Stored Products Res. *38*, 329–339.

El-Nezami, H., Mykkanen, H., Kankaanpa, P., Salminen, S., and Ahokas, J. (2000). Ability of Lactobacillus and Propionibacterium strains to remove aflatoxin B, from the chicken duodenum. J. Food Prot. *63*, 549–552.

Ehlbeck, J., Schnabel, U., Polak, M., Winter, J., von Woedtke, T., Brandenburg, R., Hagen, T., and Weltmann, K.D. (2011). Low temperature atmospheric pressure plasma sources for microbial decontamination. J. Phys. D: Appl. Phys. *44*, 13002–13008.

European Food Safety Authority (2009). Review of mycotoxin-detoxifying agents used as feed additives: mode of action, efficacy and feed/food safety. Scientific Report, EFSA, CFP/EFSA/FEEDAP/2009/01, *p*. 192.

Filatova, I., Azharonok, V., Shik, A., Antoniuk, A., and Terletskaya, N. (2012). Fungicidal effects of plasma and radio-wave pre-treatments on seeds of grain crops and legumes. NATO Sci. Peace Secur. Ser. A Chem. Biol. 469–479.

FAO (1996). Worldwide regulations for mycotoxins FAO. Food and Nutrition. Paper, vol. 55. FAO, Rome.

Fowles, J., Mitchell, J., and McGrath, H. (2001). Assessment of cancer risk from ethylene oxide residues in spices imported into New Zealand. Food Chem. Toxicol. *39*, 1055–1062.

Garofulic, I.E., Jambrak, A.R., Milosevic, S., Dragovic-Uzelac, V., Zoric, Z., and Herceg, Z. (2015). The effect of gas phase plasma treatment on the anthocyanin and phenolic acid content of sour cherry Marasca (*Prunus cerasus* var. Marasca) juice. LWT – Food Sci. Technol., 894–900.

Grenier, B., and Oswald, I.P. (2011). Mycotoxin co-contamination of food and feed; meta-analsis of publications describing toxicological interactions. W. Mycot. J. *4*, 285–313.

Grzegorzewski, F., Ehlbeck, J., Schlüter, O., Kroh, L.W., and Rohn, S. (2011). Treating lamb's lettuce with a cold plasma e Influence of atmospheric pressure Ar plasma immanent species on the phenolic profile of Valerianella locusta. LWT – Food Sci. Technol. *44*, 2285–2289.

Haidukowski, M., Pascale, M., Perrone, G., Pancaldi, D., Campagna, C., and Visconti, A. (2004). Effect of fungicides on the development of Fusarium head blight, yield and deoxynivalenol accumulation in wheat inoculated under field conditions with *Fusarium graminearum* and *Fusarium culmorum*. J. Sci. Food Agricul. *85*, 191–198.

Hertwig, C., Reineke, K., Ehlbeck, J., Erdogdu, B., Rauh, C., and Schluter, O. (2015). Impact of remote plasma treatment on natural microbial load and quality parameters of selected herbs and spices. J. Food Eng. *167*, 12–17.

Hormisch, D., Brost, I., Kohring, G.W., Giffhorn, F., Kroppenstedt, R.M., Stackebradt, E., Farber, P., and Holzapfel, W.H. (2004). *Mycobacterium fluoranthenivorans* sp nov., a fluoranthene and aflatoxin B-1 degrading bacterium from contaminated soil of a former coal gas plant. Syst. Appl. Microbiol. *27*, 653–660.

IARC (1993). Some naturally occurring substances: Food items and constituents, heterocyclic amines and mycotoxins. IARC: Lyon.

Jard, G., Liboz, T., Mathieu, F., Guyonvarc'h, A., and Lebrihi, A. (2011). Review of mycotoxin reduction in food and feed: from prevention in the field to detoxification by adsorption or transformation. Food Add. Contam. *28*, 1590–1609.

Jouany, J.P. (2007). Methods for preventing, decontaminating and minimizing the toxicity of mycotoxins in feeds. Ani. Feed Sci. Technol. *137*, 342–362.

Kang, M.H., Hong, Y.J., Attri, P., Sim, G.B., Lee, G.J., Panngom, K., Kwon, G.C., Choi, E.H., Uhm, H.S., and Park, G. (2014). Analysis of the antimicrobial effects of nonthermal plasma on fungal spores in ionic solutions. Free Radical Biol. Med. *72*, 191–199.

Khattak, A.B., and Klopfenstein, C.F. (1989). Effects of Gamma Irradiation on the Nutritional Quality of Grain and Legumes. I. Stability of Niacin, Thiamin, and Riboflavin' Cereal Chem. *66*, 169–170.

Kim, J.E., Lee, D.U., and Min, S.C. (2014). Microbial decontamination of red pepper powder by cold plasma. Food Microbiol. *38*, 128–136.

Kumar, V., Basu, M.S., and Rajendran, T.P. (2008). Mycotoxin research and mycoflora in some commercially importantagricultural commodities. Crop Protection 27, 891–905.

Kvam, E., Davis, B., Mondello, F., and Garner, A.L. (2012). Nonthermal atmospheric plasma rapidly disinfects multidrug-resistant microbes by inducing cell surface damage. Antimicrob. Agents Chemother. *56*, 2028–2036.

Lacombe, A., Niemira, B.A., Gurtler, J.B., Fan, X., Sites, J., Boyd, G., and Chen, H. (2015). Atmospheric cold plasma inactivation of aerobic microorganisms on blueberries and effects on quality attributes. Food Microbiol. *46*, 479–484.

Lee, T., Puligundla, P., and Mok, C. (2015). Inactivation of foodborne pathogens on the surfaces of different packaging materials using low-pressure air plasma. Food Control. *51*, 149–155.

Ling, L., Jiafeng, J., Jiangang, L., Minchong, S., Xin, H., Hanliang, S., and Yuanhua, D. (2014). Effects of cold plasma treatment on seed germination and seedling growth of soybean. Scientific Rep. *4*, 5859.

McKenzie, K.S., Sarr, A.B., Mayura, K., Bailey, R.H., Miller, D.R., Rogers, T.D., Norred, W.P., Voss, K.A., Plattner, R.D., Kubena, L.F., and Phillips, T.D. (1997). Oxidative degradation and detoxification of mycotoxins using a novel source of ozone. Food Chem. Toxicol. *35*, 807–820.

Magan, N., and Aldred, D. (2007). Post-harvest control strategies: Minimizing mycotoxins in the food chain. Int. J. Food Microb. *119*, 131–139.

Maragos, C.M. (2010). Zearalenone occurance and human exposure. World Myctoxin. J. *3*, 369–383.

Marasas, W.F.O. (2001). Discovery and occurrence of the fumonisins: a historical perspective. Environ. Health Pers. *109*, 239–243.

Markov, K., Mihaljevi, B., Domijan, A.M., Pleadin, J., Delas, F., and Frece, J. (2015). Inactivation of aflatoxigenic fungi and the reduction of aflatoxin B1 in vitro and in situ using gamma irradiation. Food Control. *54*, 79–85.

Masoud, W., and Kaltoft, C.H. (2006). The effects of yeasts involved in the fermentation of *Coffea arabica* in East Africa on growth and ochratoxin A (OTA) production by *Aspergillus ochraceus*. Int. J. Food Microbiol. *106*, 229–234.

Matan, N., Puangjinda, K., Phothisuwan, S., and Nisoa, M. (2015). Combined antibacterial activity of green tea extract with atmospheric radio-frequency plasma against pathogens on fresh-cut dragon fruit. Food Control *50*, 291–296.

Mendez-Albores, A., Del Rio-Garcia, J.C., and Moreno-Martinez, E. (2007). Decontamination of aflatoxin duckling feed with aqueous citric acid treatment. Animal Feed Sci. Technol. *135*, 249–262.

Miller, D.M., and Wilson, D.M. (1994). Veterinary diseases related to aflatoxin. In: The Toxicology of Aflatoxins: Human Health, Veterinary, and Agricultural Significance. Eaton, D.L. & Groopman, J.D., pp. 347–364, Academic Press, San Diego, USA.

Mishra, H.N., and Das, C. (2003). A review on biological control and metabolism of aflatoxin. Crit. Rev. Food Sci. *43*, 245–264.

Misra, N.N., Keener, K.M., Bourke, P., Mosnier, J.P., and Cullen, P.J. (2014a). In-package atmospheric pressure cold plasma treatment of cherry tomatoes. J. Biosci. Bioeng. *118*, 177–182.

Misra, N.N., Patil, S., Moiseev, T., Bourke, P., Mosnier, J.P., Keener, K.M., and Cullen, P.J. (2014b). In-package atmospheric pressure cold plasma treatment of strawberries. J. Food Eng. *125*, 131–138.

Muranyi, P., Wunderlich, J., and Heise, M. (2007). Sterilization efficiency of a cascaded dielectric barrier discharge. J. Appl. Microbiol. 103, 1535–1544.

Muranyi, P., Wunderlich, J., and Heise, M. (2008). Influence of relative gas humidity on the inactivation efficiency of a low temperature gas plasma. J. Appl. Microbiol. *104*, 1659–1666.

Na, Y., Park, G., Choi, E., and Uhm, H. (2013). Effects of the physical parameters of a microwave plasma jet on the inactivation of fungal spores. Thin Solid Films *547*, 125–131.

Newberne, P.M., Russo, R., and Wogan, G.N. (1966). Acute toxicity of aflatoxin B1 in the dog. Vet. Pathol. *3*, 331–340.

O'Brien, E., and Dietrich, D.R. (2005). Ochratoxin A:The continuing enigma. Crit. Rev. Toxicol. *35*, 33–60.

Ohkawa, H., Akitsu, T., Tsuji, M., Kimura, H., Kogoma, M., and Fukushima, K. (2006). Pulse-modulated, high-frequency plasma sterilization at atmospheric-pressure. Surf. Coat. Technol. *200*, 5829–5835.

Ouf, S.A., Basher, A.H., and Mohame, A.A. (2014). Inhibitory effect of double atmospheric pressure argon cold plasma on spores and mycotoxin production of Aspergillus niger contaminating date palm fruits. J Sci Food Agric. doi: 10.1002/jsfa.7060.

Pankaj, S.K., Misra, N.N., and Cullen, P.J. (2013). Kinetics of tomato peroxidase inactivation by atmospheric pressure cold plasma based on dielectric barrier discharge. Innov. Food Sci. Emerg. Technol. *19*, 153–157.

Park, G., Ryu, Y., Hong, Y., Choi, E., and Uhm, H. (2012). Cellular and molecular responses of *Neurospora crassa* to non-thermal plasma at atmospheric pressure. Appl. Phys. Lett. *100*, 063703–063704.

Park, B.J., Takatori, K., Konishi, Y.S., Kim, I.-H., Lee, M.-H., Han, D.-W., Chung, K.-H., Hyun, S.O., and Park, J.C. (2007). Degradation of mycotoxins using microwave-induced argon plasma at atmospheric pressure. Surface Coatings Technol. *201*, 5733–5737.

Park, D.P., Davis, K., Gilani, S., Alonzo, C.-A., Dobrynin, D., Friedman, G., Fridman, A., Rabinovich, A., and Fridman, G. (2013). Reactive nitrogen species produced in water by non-equilibrium plasma increase plant growth rate and nutritional yield. Curr. Appl. Phys. *13*, S19–S29.

Park, J.W., Lee, C., and Kim, Y.B. (2005). Fate of aflatoxin B1 during the cooking of Korean polished rice. J. Food Protect. *68*, 1431–1434.

Park, D.L. (1993). Perspectives on mycotoxin decontamination procedures. Food Add. Contaminants *10*, 49–60.

Patel, U.D., Govindarajan, P., and Dave, P.J. (1989). Inactivation of aflatoxin B1 by using the synergistic effect of hydrogen peroxide and gamma radiation. Appl Environ Microbiol. *55*, 465–467.

Pereyra, M.L.G., Carvalho, E.C.Q., Tissera, J.L., Keller, K.M., Magnoli, C.E., Rosa, C.A.R., Dalcero, A.M., and Cavaglieri, L.R. (2008). An outbreak of acute aflatoxicosis on a chinchilla (Chinchilla lanigera) farm in Argentina. Case Reports. J. Vet. Diag. Invest. *20*, 853–856.

Perni, S., Shama, G., and Kong, M.G. (2008). Cold atmospheric plasma disinfection of cut fruit surfaces contaminated with migrating microorganisms. J. Food Prot. *71*, 1619–1625.

Pestka, J.J., and Smolinski, A.T. (2005). Deoxynivalenol: toxicology and potential effects on humans. J. Toxicol. Environ. Health, Part B: Crit. Rev. *8*, 39–69.

Pfohl-Leszkowicz, A., Chekir-Ghedira, L., and Bacha, H. (1995). Genotoxicity of zearalenone, an estrogenic mycotoxin: DNA adduct formation in female mouse tissues. Carcinogenesis *16*, 2315–2320.

Richard, J.L. (2007). Some major mycotoxins and their mycotoxicoses – an overview. Int. J. Food Microbiol. *119*, 3–10.

Rico, D., Martin-Diana, A., Barat, J.M., and Barry-Ryan, C. (2007). Extending and measuring the quality of fresh-cut fruit and vegetables: a review, Trends Food Sci. Technol. *18*, 373–386.

Ritieni, A., Monti, S.M., Moretti, A., Logrieco, A., Gallo, M., Ferracane, R., and Fogliano, V. (1999). Stability of fusaproliferin, a mycotoxin from *Fusarium* spp. J. Sci. Food Agric. *79*, 1676–1680.

Roll, R., Matthiaschk, G., and Korte, A. (1990). Embryotoxicity and mutagenicity of mycotoxins. J. Environ. Pathol. Toxicol. Oncol. *10*, 1–7.

Scholtz, V., Pazlarová, J., Sousková, H., Khun, J., and Julak, J. (2015). Nonthermal plasma a tool for decontamination and disinfection. Biotech. Advances. doi:10.1016/j.biotechadv.2015.01.002

Selcuk, M., Oksuz, L., and Basaran, P. (2008). Decontamination of grains and legumes infected with *Aspergillus* spp. and *Penicillum* spp. by cold plasma treatment. Biores. Technol. *99*, 5104–5109.

Shapira, R., and Paster, N. (2004). Control of mycotoxins in storage and techniques for their decontamination. In Magan, N., and Olsen, M. (Eds.), Mycotoxins in food. Cambridge: Woodhead Publishing Ltd and CRC Press LLC., pp. 196–198.

Sharma, R.P. (1993). Immunotoxicity of mycotoxins. J. Dairy Sci. *76*, 892–897.

Shintani, H., Sakudo, A., Burke, P., and McDonnell, G. (2010). Gas plasma sterilization of microorganisms and mechanisms of action. Exp. Ther. Med. *1*, 731–738.

Steiman, R., Seigle-Murandi, F., Sage, L., and Krivobok, S. (1989). Production of patulin by micromycetes. Mycopathologia *105*, 129–133.

Subirade, I. (1996). Fate of ochratoxin A during breadmaking. Food Addit. Contam. *13*, 25–26.

Suhem, K., Matan, N., Nisoa, M., and Matan, N. (2013). Inhibition of *Aspergillus flavus* on agar media and brown rice cereal bars using coldatmospheric plasma treatment. Int. J. Food Microbiol. *161*, 107–111.

Surowsky, B., Fröhling, A., Gottschalk, N., Schlüter. O., and Knorr, D. (2014). Impact of cold plasma on *Citrobacter freundii* in apple juice: inactivation kinetics and mechanisms. Int. J. Food Microbiol. *174*, 63–71.

Takai, E., Kitano, K., Kuwabara, J., and Shiraki, K. (2012). Protein Inactivation by low-temperature atmospheric pressure plasma in aqueous solution. Plasma Proces. Poly. *9*, 77–82.

Tolleson, W.H., Melchior, W.B., Morris, S.M., McGarrity, L.J., Domon, O.E., Muskhelishvili, L., James, S.J., and Howard, P.C. (1996). Apoptotic and anti-proliferative effects of fumonisin B1 in human keratinocytes, fibroblasts, esophageal epithelial cells and hepatoma cells. Carcinogenesis *17*, 239–249.

Varga, J., Kevei, E., Rinyu, E., Téren, J., and Kozakiewicz, Z. (1996). Ochratoxin production by Aspergillus species. App.l Environ. Microbiol. *62*, 4461–4464.

Velluti, A., Sanchis, V., Ramos, A.J., Egido, J., and Marin, S. (2003). Inhibitory effect of cinnamon, clove, lemongrass, oregano and palmarose essential oils on growth and fumonisin B1 production by *Fusarium proliferatum* in maize grain. Int. J. Food Microbiol. *89*, 145–154.

Volin, J.C., Denes, F.S., Young, R.A., and Park, S.M.T. (2000). Modification of seed germination performance through cold plasma chemistry technology. Crop Sci. *40*, 1706–1718.

Wang, S.Q., Huang, G.Q., Li, Y.P., Xiao, J.X., Zhang, Y., and Jiang, W.L. (2015). Degradation of aflatoxin B1 by lowtemperature radio frequency plasma and degradation product elucidation. Eur. Food Res. Technol. DOI 10.1007/s00217-015-2439-5.

Ward, T.J., Bielawski, J.P., Kistler, H.C., Sullivan, E., and O'Donnell, K. (2002). Ancestral polymorphism and adaptative evolution in the trichothecene mycotoxin gene cluster of phytopathogenic *Fusarium*. Proc. Natl. Acad. Sci. U.S.A. *99*, 9278–9283.

Wild, C.P., and Turner, P.C. (2002). The toxicology of aflatoxins as a basis for public health decisions. Mutagenesis *17*, 471–481.

Williams, J.H., Phillips, T.D., Jolly, P.E., Stiles, J.K., Jolly, C.M., and Aggarwal, D. (2004). Human aflatoxicosis in developing countries: a review of toxicology, exposure, potential health consequences, and interventions. Am. J. Clin. Nutr. *80*, 1106–1122.

Winkler, J., Kersten, S., Meyer, U., Engelhardt, U., and Dänicke, S. (2013). Residues of zearalenone (ZEN), deoxynivalenol (DON) and their metabolites in plasma of dairy cows fed Fusarium contaminated maize and their relationships to performance parameters. Food Chem. Toxicol. *65*, 196–204.

World Health Organization (WHO) (1999). Basic Food Safety for HealthWorkers. WHO: Geneva.

Yang, L.Q., Chen, J.R., Gao, J.L., and Guo, Y.F. (2009). Plasma sterilization using the RF glow discharge. Appl. Surf. Sci. *255*, 8960–8964.

Yazdanpanah, H., Mohammadi, T., Abouhassani, G., and Cheraghali, A.M. (2005). Effect of roasting on degradation of aflatoxins in contaminated pistachio nuts. Food, Chem. Toxicol. *43*, 1135–1139.

Yin, Y.N., Yan, L.Y., Jiang, J.H., and Ma, Z.H. (2008). Biological control of aflatoxin contamination of crops. J. Zhejiang Univ. Sci B. *9*, 787–792.

Yoshizawa, T., Yamashita, A., and Luo, Y. (1994). Fumonisin occurrence in maize from high- and low-risk areas for human esophageal cancer in China. Appl. Environ. Microbiol. *60*, 1626–1629.

Yu, J.Hy *Staphylococcus aureus*: effects and mechanisms. Plasma Chem. Plasma Proces. *29*, 229–240.

Zhu, R., Feussner, K., Wu, T., Yan, F., Karlovsky, P., Zheng. X. (2015a) Detoxification of mycotoxin patulin by the yeast *Rhodosporidium paludigenum*. Food Chem. *179*, 1–5.

Zhu, C., Shi, J., Jiang, C., and Liu, Y. (2015b). Inhibition of the growth and ochratoxin A production by *Aspergillus carbonarius* and *Aspergillus ochraceus* in vitro and in vivo through antagonistic yeasts. Food Control *50*, 125–132.

Current Progress in the Sterilization of Spores and Vegetative Cells by Exposure to Gas Plasma: Sterilization, Disinfection and Antimicrobial Activity

8

Hideharu Shintani

Abstract

In general, spores are more tolerant than vegetative cells against sterilization and disinfection. Thus, spores are generally used as the biological indicator (BI), which should correspond to the microorganism most tolerant to the targeted sterilization process (ISO 11138-1, ISO 14161). *Geobacillus stearothermophilus* ATCC 7953 was therefore selected as the BI for gas plasma sterilization because its spores are the most tolerant of the tested microorganisms. In this chapter, the relationship between BI and ISO, and their importance for validation of sterilization are described in detail.

Introduction

Sterilization is the most stringent process to eliminate microorganisms (see Table 2.2) (Shintani and McDonnell, 2011; Shintani *et al.*, 2007). In this chapter sterilization of spores and vegetative cells is described (Box 8.1). According to Table 2.2, spores are the most tolerant form of microorganisms. Prions are not microorganisms but rather are proteins, so prions will not be discussed in this chapter although they are more difficult to eliminate than spores (see Table 2.2). The killing of spores and other types of microorganisms including vegetative cells is called sterilization (Shintani and McDonnell, 2011; Magureanu *et al.*, 2011). Disinfection refers to the process of killing vegetative cells, but not spores (Shintani and McDonnell, 2011; Hoffmann *et al.*, 2013), and decontamination is the removal of microorganisms by processes that do not necessarily kill the microorganisms (Shintani and McDonnell, 2011).

Sterilization validation study using BI or bioburden

Under real sterilization conditions, the real target of killing is the bioburden, not the BI. The bioburden is defined the number and type of viable microorganisms on/in the products (ISO 14161, ISO 11138-1, ISO 14937). Spores are relatively rare in the environment, and

Box 8.1 List of publications where sterilization of spores and vegetative cells is described

Shintani *et al.*, 2010; Tseng *et al.*, 2012; Shi *et al.*, 2009; Roth *et al.*, 2010: Hong *et al.*, 2009; Lee *et al.*, 2006; Lassen *et al.*, 2006; Lassen *et al.*, 2005; Purevdorj *et al.*, 2003; Lerouge *et al.*, 1999; Lerouge *et al.*, 2000; Hury *et al.*, 1998; Patil *et al.*, 2014: Stapelmann *et al.*, 2013: Kawamura *et al.*, 2012: Colas *et al.*, 2012: Muranyi *et al.*, 2010; Ayliffe *et al.*, 2000; Mols *et al.*, 2013; Joaquin *et al.*, 2009; Winter *et al.*, 2013; Alfa *et al.*, 1996; Baier *et al.*, 1992; Sakudo *et al.*, 2013: Fricke *et al.*, 2012; Brun *et al.*, 2012; Kvam *et al.*, 2012; Liu *et al.*, 2011; Brelles-Marino *et al.*, 2012; Rederstorff *et al.*, 2011; Hauser *et al.*, 2011; Rainer *et al.*, 2010; Klnnari *et al.*, 2010; Muranyi *et al.*, 2008; Rowan *et al.*, 2008; Muranyi *et al.*, 2007; Tessarolo *et al.*, 2006; Yu *et al.*, 2007; Peniston and Choi, 2007; Donohue *et al.*, 2006; Markowicz *et al.*, 2006; Baxter *et al.*, 2005; Stoffels, *et al.*, 2004; Whittaker *et al.*, 2004; Lassen *et al.*, 2003; Hopper *et al.*, 2003; Trostie *et al.*, 2002; Shimizu *et al.*, 2001; Moisan *et al.*, 2001; Ferreira *et al.*, 2001; Drews, 2001; Liao *et al.*, 2001; Duffy *et al.*, 2000; Smith *et al.*, 2000; Reeves *et al.*, 2000; Adler *et al.*, 1998; Bryce *et al.*, 1997; Graham and Mieinik, 1997; Fisher *et al.*, 1997; Hesby *et al.*, 1997; Crow and Smith, 1995; Grriffiths, 1983; Geiss, 1994; Draeger *et al.*, 1990; Ehrman, 1989; Anderson, 1989; Hosseinzaden *et al.*, 2013; Hayes *et al.*, 2013; Abreu *et al.*, 2013; Noriega *et al.*, 2011; Cotter *et al.*, 2011; Haffmann *et al.*, 2013; Wiegand *et al.*, 2014; Bayliss *et al.*, 2013; Matthes *et al.*, 2012; Niemira *et al.*, 2014; Niemira, 2012; Niemira, 2011; Ermolaeva *et al.*, 2012; Ermolaeva *et al.*, 2011; Fan *et al.*, 2011; Vandamme *et al.*, 2012; Kim *et al.*, 2010; Bai *et al.*, 2009; Burgassi *et al.*, 2009; Du *et al.*, 2008; Silva *et al.*, 2007; Luo and Lehotay, 1997; Jungersten *et al.*, 1996; Rubbo *et al.*, 1995; Griffiths, 1993; Matthes *et al.*, 2014; Edelblute *et al.*, 2014; Perni *et al.*, 2008; Rowan *et al.*, 2007; Alkawareek *et al.*, 2012; Mishra *et al.*, 2003; Trosite *et al.*, 2002; Brown *et al.*, 2002; Pignata *et al.*, 2014; Mois *et al.*, 2013; Fernandez *et al.*, 2013; Winter *et al.*, 2011; Whittaker *et al.*, 2004; Yamamoto, 1999; Hosseninzadeh *et al.*, 2013; Vandervoort and Brelles-Marino, 2014; Nisshioka *et al.*, 2014; Kim *et al.*, 2014; Yong *et al.*, 2014; Lackmann and Bandow, 2014; Zhang *et al.*, 2014; Han *et al.*, 2014; Al-Mariri *et al.*, 2013; de Oliveira Cardoso Macedo, M *et al.*, 2013; Hosseinzadeh Colagar *et al.*, 2013; Hernandez-Arias *et al.*, 2012; Lee *et al.*, 2011; Fernandez *et al.*, 2011; Kim *et al.*, 2011; Yu *et al.*, 2006; Ortoneda *et al.*, 2008; Kim *et al.*, 2014; Harkln *et al.*, 2001; Wang *et al.*, 2014; Vetten *et al.*, 2014; Delgado *et al.*, 2014; Justan *et al.*, 2014; Wittenburg *et al.*, 2014; Traba *et al.*, 2013; Sakudo *et al.*, 2013; Thiyagarajan *et al.*, 2013; Pokorny *et al.*, 2012; Lee and Chol, 2012; Deeilmann *et al.*, 2008; Hauser *et al.*, 2008; Lleixa *et al.*, 2008; Selcuk *et al.*, 2008; Kang *et al.*, 2014; Boscariol *et al.*, 2008; Tank *et al.*, 2013; Southwood and Baxter, 1996; Zhao and Lubman, 1993; Chen *et al.*, 2010; Guo *et al.*, 2008; McCullough and Bartfay, 2007; Sun *et al.*, 2013; Nishioka *et al.*, 2014; Manolache *et al.*, 2001; Aerts *et al.*, 2013; MacDonald *et al.*, 2012; Yuan *et al.*, 2010; Liang *et al.*, 2010; Deilmann *et al.*, 2008; Alfadhli *et al.*, 2008; Shen *et al.*, 2008; Trostle *et al.*, 2002; Collier *et al.*, 1996; Dellmann *et al.*, 2008; Zhamg *et al.*, 2014; Lupu and Georgescu, 2010; Schultz-Jensen *et al.*, 2011; Wang *et al.*, 2008; Kanemitsu *et al.*, 2005; Yuen *et al.*, 2011; Isbary, 2013; Idlibi *et al.*, 2013; Gatineau *et al.*, 2012; Matthes *et al.*, 2010; Yang *et al.*, 2010; Gao *et al.*, 2006; Zhang *et al.*, 2004; Yet-Pole *et al.*, 2004; Alastuey and Ballenegger, 2012; Frazer, 1999.

the bioburden contaminating most products is generally composed of vegetative cells (Table 8.1). Typical examples of airborne, falling and adhesive microorganisms at a healthcare facility are presented in Tables 8.1–8.3 (Shintani *et al.*, 2004, 2006). The CFU (colony-forming unit) of airborne microorganisms was around 10 CFU/500 L (Shintani *et al.*, 2004, 2006). According to ISO 14161 and sterilization validation, reduction of 10^6 CFU of spores to a sterility assurance level (SAL) of 10^{-6} is required in sterilization validation in the case of the overkill method in ISO 14161. In ISO 14161, an initial population approximating the number of the bioburden or 10^3 CFU/carrier is also approved (ISO 14161), but this will be discussed in detail in Chapter 13.

The requirement of an initial population of 10^6 CFU/carrier and a SAL of 10^{-6} in ISO 11138-1 is specifically for BI manufacturers carrying out validation studies. In contrast, an initial population of 10^6 CFU/carrier is normally unnecessary for BI users (ISO 14161), but a SAL of 10^{-6} is definitely required for both BI manufacturers (ISO 11138-1) and BI users

Table 8.1 Airborne microorganisms identified in the Namiki Clinic dialysis room

Bacterial species	CFU
Staphylococcus haemolyticus	3
Staphylococcus hominis	2
Staphylococcus schleiferi	2
Staphylococcus epidermidis	7
Staphylococcus intermedius	1
Staphylococcus saprophyticus	2
Staphylococcus capitis	1
Staphylococcus cohnii subsp. *cohnii*	1
Staphylococcus pasteuri	1
Staphylococcus vitulus	1
Streptococcus sanguinis	1
Micrococcus luteus	6
Micrococcus sedentarius	6
Micrococcus species	4
Bacillus licheniformis	4
Bacillus subtilis	7
Bacillus megaterium	1
Acinetobacter lwoffii	2
Lactobacillus raffinolactis	1
Actinomyces pyogenes	3
Saccharomyces species	1
Corynebacterium genitalium	1
Gardnerella vaginalis	1
Pantoea agglomerans	1

CFU, colony-forming units. Data for fungi, moulds and yeast other than *Saccharomyces* are omitted.

Table 8.2 Number of colonies after incubation at 30°C

Medium	Equipment	Date						Total	Average
		Jul 9	Jul 10	July 11	July 12	July 13	July 16		
		Sum CFU of $n=3$							
SCDA	A	3	16	9	15	16	19	78	4.3
	B	10	14	9	10	27	26	96	5.3
	C	10	11	11	10	18	3	63	3.5
	D	10	15	6	7	19	3	60	3.3
	E	9	21	6	5	9	27	77	4.3
SCDALP	A	15	17	17	16	20	14	99	5.5
	B	23	17	12	21	15	21	109	6.1
	C	23	12	16	17	10	21	99	5.5
	D	15	11	13	13	27	12	99	5.5
	E	8	12	9	14	15	18	72	4.0

30°C incubation in SCDA or SCDALP indicates bacterial cultivation.
SCDA: soybean casein digest agar
SCDLP: soybean casein digest agar lecithin polysorbate
Polysorbate indicates Tween[R] as trademark.

Table 8.3 Number of colony-forming units (total of three samples) after incubation at 20°C

Medium	Equipment	Date						Total	Average
		9 July	10 July	11 July	12 July	13 July	16 July		
SCDA	A	2	23	7	9	14	10	65	3.6
	B	3	15	13	16	11	17	75	4.2
	C	4	8	4	5	3	1	25	1.4
	D	3	12	4	12	20	7	58	3.2
	E	8	8	4	6	12	6	44	2.4
SCDALP	A	5	13	5	11	17	10	51	3.4
	B	14	11	11	16	7	24	83	4.6
	C	11	3	6	18	7	8	53	2.9
	D	5	6	5	7	15	12	50	2.8
	E	12	12	5	14	10	18	71	3.9

20°C incubation in SCDA or SCDALP indicates fungi, yeast and mould cultivation.
SCDA, soybean casein digest agar; SCDL, soybean casein digest agar lecithin polysorbate.
Polysorbate indicates Tween[R] as trademark.

(ISO 14161). A population of 10^6 CFU/carrier is defined as the smallest population that is visibly turbid. A SAL of 10^{-6} can be defined as the closest number to zero from the stochatics in ISO 11737-1.

When an initial population of 10^6 CFU is used, sterilization must result in a SAL of 10^{-6}, indicating that a 12 log reduction is required in the case of BI manufacturers, which is an absolute requirement. However, this is not an absolute requirement for BI users.

ISO documents do not specify a particular biological indicator (BI) for use in gas plasma sterilization, but according to our experiments (Shintani *et al.*, 2007) and those of others, *Geobacillus stearothermophilus* ATCC 7953 is more tolerant to sterilization than *Bacillus atrophaeus* ATCC 9372 (Lasser *et al.*, 2006; Klaempfl *et al.*, 2012), so we decided to use *Geobacillus stearothermophilus* ATCC 7953 as the BI for gas plasma sterilization. According to the definition of BI in ISO 11138-1, the most sterilization-resistant non-pathogenic microorganism should be selected as the BI. Therefore the use of *Geobacillus stearothermophilus* ATCC 7953 as the BI is appropriate (ISO 11138-1, ISO 14161). Deng *et al.* (2006) selected *Bacillus subtilis* in place of *Geobacillus stearothermophilus* ATCC 7953, but there is no rationale for using this type of sporeformer as the BI.

Gas plasma sterilization of *Geobacillus stearothermophilus* ATCC 7953 and *Bacillus atrophaeus* ATCC 9372 spores resulted in linear survivor curves with no tailing (see Fig. 1.4), indicating first order inactivation kinetics. The *D* value (decimal reduction value, time or dose to reduce 1 log) can be calculated from this straight survivor curve (ISO 14161). For example, based on data presented in Fig. 1.4, it takes 7 min for a 6 log reduction, indicating a *D* value of 1.2 min. It is better to avoid using the fraction negative procedures of Stumbo–Murphy–Cochran and Sperman–Karber (ISO 14161) when utilizing gas plasma sterilization because the survivor curve method results in a more precise *D* value. Fraction negative methods require fewer numbers of BI, but they require a premise that from the fraction negative range (SAL 5 to SAL 10^{-2}) to the initial population of 10^6 CFU/carrier must be a straight line (Fig. 3.1). It is impossible to confirm experimentally whether or not the line is straight for a SAL of less than 10^{-2}. The reduction from 10^6 CFU/carrier to SAL of 10^{-2} must be confirmed experimentally to be linear and then reduction from a SAL of 10^{-2} to 10^{-6} can be speculated to be linear (Fig. 3.1). The survivor curves of the overkill and half cycle methods in ISO 14161 are required to be linear from an initial population of 10^6 CFU/carrier to a SAL of 10^{-6}. This topic will be discussed in detail in Chapter 13.

The *D* value is only one per one microorganism. When a tailing phenomenon occurs, there is more than one *D* value, which is a serious error (Rossi and Kylian, 2012).

In real world situations the real target of the sterilization process is not always spores, but also primarily vegetative cells. The bioburden consists primarily of vegetative cells; spores are relatively rare (Table 8.1). The *D* value of vegetative cells is ~1/50 to 1/200 that of spores, indicating that we are required to carry out an unnecessarily excessive time/dose exposure to achieve sterilization validation according to the authorities. In real situations, the population of the bioburden is typically only a few CFU (10^0 CFU level) and clumping or multilayer phenomena (Fig. 1.1) are rarely observed. In spite of this low level of contamination in typical situations, in sterilization validation studies we are required to use 10^6 CFU of *Geobacillus stearothermophilus* ATCC 7953 spores as an initial population and must achieve a SAL of 10^{-6} as indicated in ISO 11138-1 and 14161. A SAL of 10^{-6} is required for sterilization, disinfection, or decontamination, which is correct, but an initial population of 10^6 spore/carrier is an excessive and unrealistic requirement. Such an excessive requirement is likely to result in a failure to attain material and functional compatibility. As already mentioned, it is quite important to attain both a SAL of 10^{-6} and material/functional compatibility in Chapter 3. To attain a SAL of 10^{-6} from the initial population of 10^6 CFU/carrier is a full 12 log reduction, which is required of BI manufacturers in validation studies (ISO 11138-1). However, for the BI manufacturer, there is no requirement of simultaneous achievement of material/functional compatibility (ISO 11138-1) because there is no material being tested. Therefore, this means that the requirements for validation studies carried out by BI users (ISO 14161) and BI manufacturers (ISO 11138-1 differ significantly.

The concept of an initial population of 10^6 CFU/carrier is the least visible number seen as turbidity and a SAL of 10^{-6} is the closest number to zero calculated from stochastics (ISO 11137-1). In sterilization, disinfection and decontamination, a SAL of 10^{-6} may be essential, but an initial population of 10^6 CFU of spores is not realistic. In real sterilization situations, the bioburden is mostly vegetative cells (Table 8.1), the number of bioburden on the products is only a few, and no clumping (multilayer) is observed. If the initial population is a few CFU (10^0 CFU level), then the requirement of a SAL of 10^{-6} means a 6 log reduction. Therefore, a 6 log reduction is the actual decrease required for BI users. Most engineering researchers mistake an initial population of 10^6 CFU/carrier to a 6 log reduction means 0 (total death) as the actual requirement, which is incorrect. A 6 log reduction of an initial population of 10^6 CFU/carrier is $10^0 = 1$, not zero. A SAL of 10^0 means there is a 63% possibility of survival (POS; see Fig. 3.1). The exact six log reduction required of the BI user is from an initial population of 10^0 CFU (bioburden level) to a SAL of 10^{-6}. This is the real six log reduction required of BI users.

A 6 log reduction may be more realistic, but there is no commercially available 10^0 to 10^2 CFU/carrier BI. Commercially available BIs are from 10^3 CFU/carrier, which are utilized in the combined BI/bioburden method in sterilization validation (ISO 14161). The combined BI/bioburden method can be used with an initial population of BI approximately equivalent in number to that of the bioburden, but this concentration of BI is not commercially available and must be self-made. When the BI is made by the user, this requires that ISO 11138-1 must be followed, as the BI user is temporarily regarded as the BI manufacturer in this case. In place of using the bioburden number of BI, a commercially available BI with $> 10^3$ CFU/carrier is approved for use in sterilization validation according to ISO 14161. For example, when 10^3 CFU/carrier as an initial population was used and attained SAL of 10^{-6}, a 9 log reduction is required (ISO 14161). However, many inspectors do not know these alternative requirements, and mistakenly assume that 10^6 CFU/carrier must be used as the initial population. Therefore, the level of 10^6 CFU/carrier is unchanged due to this misconception and it remains the most commercially popular BI. It is correct for the BI manufacturer to use an initial population of 10^6 CFU/carrier in validation studies as described in ISO 11138-1, not for BI users. BI users do not need to obey ISO 11138-1 unless they are making their own BI; rather, they must obey ISO 14161. In ISO 14161 the absolute bioburden method and combined BI/bioburden method are described. These methods will be described in detail in Chapter 13.

According to our experiments (Shintani *et al.*, 2004, 2006), airborne microorganisms at a healthcare facility were present at 10–20 CFU/500L (Tables 8.1–8.3), indicating just a few CFU of falling microorganisms or a few CFU of adhesive microorganisms. Adhesive microorganisms are equal in concept to the bioburden. Even in the hospital environment, for example the dialysis room, the bioburden is a few CFU; production facilities must have strict controls to prevent contamination and there is no possibility that products will be contaminated with 10^6 CFU spore/carrier. Therefore, we request the authorities to revise the requirements such that they resemble the real world situation and consider that a 10^6 CFU/carrier requirement as an initial population is unnecessary for BI users (Healthcare product companies, etc.).

Difference of spore appearance after sterilization using several sorts of gases

In the current study of gas plasma sterilization, several types of gases are used. Typical gases are nitrogen, oxygen, argon, helium, air and so on. Oxygen gas plasma causes significant etching and shrinkage (Rossi *et al.*, 2008; Rossi and Kylian, 2012), so material/functional compatibility is difficult to attain with oxygen gas plasma. This means that oxygen gas plasma sterilization is not an appropriate procedure for practical use. Nitrogen, argon and helium do not produce any change in appearance after sterilization (Rossi *et al.*, 2008; Lerouge *et al.*, 2000), but it is not easy to attain a SAL of 10^{-6} in a short period. In order to avoid material deterioration and to attain a SAL of 10^{-6} in a relatively short period, a mixture of nitrogen and oxygen (4:1 v:v) is generally used in European countries. With a 20% of oxygen mixture, some etching is observed but it is not significant and does not cause material incompatibility, so in European countries, a 4:1 (v:v) mixture of oxygen and nitrogen is most popular. Nitrogen can be replaced with argon or helium, which are more expensive gases, so from an economical standpoint, nitrogen may be most appropriate. Nitrogen, argon and helium have the same efficiency and results of the spore appearance are identical (Rossi *et al.*, 2008; Lerouge *et al.*, 2000).

Preparation of survivor curve

Methods for the preparation of survivor curves can be found in ISO 11138-1 Normative Annex B, and those for retrieval of spores or bioburden from carrier material or from products can be found in ISO 11737-1. A detailed description can be found in Chapter 13. To prepare survivor curves, 10-fold dilutions must be repeated several times to attain 30–300 CFU/plate (ISO 11138-1); soybean casein digest agar (SCDA) plates are used (ISO 11737–1). In contrast, soybean casein digest broth (SCDB) can be used and successful sterilization can be determined based on the turbidity and change of colour of chemical indicator (CI). However, when using SCDB, colonies cannot be counted, so survivor curves cannot be generated. Due to the presence of the CI, a colour change of SCDB medium indicates that some spores survived and produced organic acids via the TCA cycle (tricarboxylic acid cycle, citric acid cycle) and these organic acids, primarily citric acid, lower the pH of the broth. However, it is important to note that CIs are not approved for use in validation studies (ISO 11138-1 and ISO 14161), so the use of SCDB for sterilization validation is invalid.

Conclusion

Use of a BI with an initial population of 10^6 CFU/carrier is not correct for BI users because this is a requirement in ISO 11138-1 for BI manufacturers carrying out validation studies, and BI users must obey ISO 14161, which describes other methods with different requirements.

A SAL of 10^{-6} must be obtained in all sterilization procedures without exception. For this purpose, any tailing phenomenon in the survivor curve must be avoided. The reason for tailing is due to clumping of the BI (see Fig. 1.1), so BI free from clumping must be used for gas plasma sterilization (Fig. 1.2). As the penetration depth of gas plasma is quite shallow (10–20 nm), gas plasma sterilization can often result in tailing. BI free from clumping is therefore absolutely required; for this purpose the author recommends BI from Merck Co.

Ltd. As an example, data in Fig. 1.4 were obtained using Merck BI (Fig. 1.2, free from clumping), and no tailing was observed. Purchased BI must be observed by scanning electron microscopy (SEM) to be free from clumping by the BI user. This is the user's responsibility even if it is not described in ISO 14161.

For gas plasma sterilization, the recommended BI to use is *Geobacillus stearothermophilus* ATCC 7953, but the spores of this organism are not defined as the official BI in any ISO documents in ISO TC 198. According to the author's experiments, *Geobacillus stearothermophilus* ATCC 7953 shows the highest tolerance to gas plasma sterilization; therefore, it is appropriate for use as the BI for gas plasma sterilization because the BI should be the most sterilization–tolerant non-pathogenic microorganism available. Based on this definition, *Geobacillus stearothermophilus* ATCC 7953 was selected as the BI for gas plasma sterilization. To compare the *D* value data among published studies it is necessary to use an identical BI from the same manufacturer if possible, because variations in the populations and *D* values of BI can be significant (Shintani and Akers, 2000; Shintani *et al.*, 2000; Shintani, 1996 1997). In ISO 1138-1, the approved variation range of the population is −50 to +300% and that of the *D* value is 1.5–2.5 min or 2.5–3.5 min. The actual variations are beyond this range (Shintani and Akers, 2000; Shintani, *et al.*, 2000; Shintani, 1996, 1997). Details will be discussed in Chapter 13.

Oxygen gas plasma causes significant shrinkage after sterilization, indicating that material compatibility is difficult to attain when using oxygen gas plasma. In place of oxygen, air gas plasma, which can easily attain a SAL of 10^{-6} and material/functional compatibility, is popular in European countries. Nitrogen, argon or helium gas plasma can also be used to attain both a SAL of 10^{-6} and material/functional compatibility. The problem with the use of these gases is the high gas prices.

References

Abreu, A.C., Tavares, R.R., Borges, A., Mergulhaeo, F., and Simoees, M. (2013). Current and emergent starategies for disinfection of hospital environments. J. Antimicrob. Chemother. *68*, 2718–3732.

Adler, S., Scherrer, M., and Daschner, F.D. (1998). Costs of low temperature plasma sterilization compared with other sterilization methods. J. Hosp. Infect. *40*, 125–134.

Aerts, R., Tu, X., Van Gaens, W., Whitehead, J.C., and Bogaerts, A. (2013). Gas purification by nonthermal plasma: a case study of ethylene. Environment. Sci. Technol. *47*, 64758–66485.

Al-Mariri, A., Saloum, S., Mrad, O., Swied, G., and Alkhaled, B. (2013). Inactivation of Gram-negative bacteria by low pressure RF remote plasma excited in N_2–O_2 mixture and SF6 gases. Iran J. Med. Sci. *38*, 334–338.

Alastuey, A., and Ballenegger, V. (2012). Thermodynamics of atomic and ionized hydrogen: analytical results versus equation-of-state tables and Monte Carlo data. Phys. RTers. Stat Nonlin Soft Matter Phys. 2012 Dec:86(6 Pt2);066402.

Alfa, M.J., DeGange, P., Olson, N., and Puchlski, T. (1996). Comparison of ion plasma, vaporized hydrogen peroxide, and 100% ethylene oxide sterilizers to the 12/88 ethylene oxide gas sterilizer. Infect. Control Hosp. Epidemiol. *17*, 92–100.

Alfadhli, A.A., Alazmi, W.M., Ponich, T., Howard, J.M., Prokoplw, I., Alaqeel, A., and Gregor, J.C. (2008). Efficacy of argon plasma coagulation compared to topical formalin application for chronic radiation proctopathy. Can. J. Gastroenterol. *22*, 129–132.

Alkawareek, M.Y., Algwari, Q.T., Gorman, S.P., Graham, W.G., O'Connell, D., and Gilmore, B.F. (2012). Application of atmospheric pressure nonthermal plasma for the in vitro eradication ofbacterial biofilms. FEBS Immunol. Med. Microbiol. *65*, 381–384.

Anderson, M.H. (1989). Gas plasma sterilization:innovation in practice. Navy Med. *80*, 9–10.

Ayliffe, G. (2000). Decontamination of minimally invasive surgical endoscopes and accessories. J. Hosp. Infect. *45*, 263–277.

Bai, Y., Chan, J., Mu, H., Zhang, C., and Li, B. (2009). Reduction of dichlorvos and omethoate residues by O_2 plasma treatment. J. Agric. Food Chem. 57, 6238–6245.

Baier, R.E., Carter, J.M., Sorensen, S.E., Meyer, A.E., McGowan, B.D., and Kapsrzak, S.A. (1992). Radiofrequency gas plasma (glow discharge) disinfection of dental operative instruments, including handpieces. J. Oral. Implant. *18*, 236–242.

Baxter, H.C., Campbell, G.A., Whittaker, A.G., Jones, A.C., Altken, A., Simpson, A.H., Casey, M., Bountiff, L., Gibbard, L., and Baxter, R.L. (2005). Elimination of transmissible spongiform encephalopathy infectivity and decontamination of surgical instrument by using radio-frequency gas-plasma treatment. J. Gen. Virol. *86*, 2393–2399.

Bayliss, D.L., Shama, G., and Kong, M.G. (2013). Restoration of antibiotic sensitivity in meticillin-resistant *Staphylococcus aureus* following treatment with a non-thermal atmospheric gas plasma. Int. J. Antimicrob. Agents *41*, 398–399.

Boscariol, M.R., Moreira, A.J., Mansano, R.D., Kikuchi, I.S., and Pinto, T.J. (2008). Sterilization by pure oxygen plasma and by oxygen-hydrogen peroxide plasma; an efficiency study. Int. J. Pharm. *353*, 170–175,

Brelles-Marino, G. (2012). Induction of a viable-but-non-culturable state in bacteria treated with gas discharge plasma. J. Appl. Microbiol. *112*, 412–413.

Brown, S.A., Merritt, K., Woods, T.O., McNamee, S.G., and Hichins, V.M. (2002). Effect of different disinfection and sterilization methods on tensile strength of materials used for single-use devices. Biomed. Instrum. Technol. *36*, 23–27.

Brun, P., Brun, P., Vono, M., Venier, P., Tarricone, E., Deligianni, V., Martines, E., Zuin, M., Spagnolo, S., Cavazzana, R., Cardin, R., Castagliuolo, I., Valerio, A.L., and Leonardi, A. (2012). Disinfection of ocular cells and tissues by atmospheric-pressure cold plasma. PLoS One 7:e33245.doi;10.1371

Bryce, E.A., Chla, E., Logelin, G., and Smith, J.A. (1997). An evaluation of the AbTox Plazlyte Sterilization system. Infect. Control. Hosp. Epidemiol. *18*, 646–653.

Burgassi, S., Zanardi, I., Travagli, V., Montomoli, E., and Bocci, V. (2009). How much ozone bactericidal activity is compromised by plasma components? J. Appl. Microbiol. *106*, 1715–1721.

Chen, H., Chen, Z.J., Wu, M.B., and Deng, S.X. (2010). Screening the Fusarium graminearum inhibitory mutant strain from Bacillus subtilis by atmospheric-pressure plasma jet. J. Appl. Microbiol. *108*, 96–103.

Colas, N.A., Espinasse, V., Perrier-Cornet, J.M., and Gervasis, P. (2012). High gas pressure; an innovative methods for the inactivation of dried bacterial spores. Biotecnol. Bioeng. *109*, 1996–2004.

Collier, J.P., Sutula, L.C., Currier, B.H., Currier, J.H., Wooding, R.E., Williams, I.R., Farber, K.B., and Mayor, M.B. (1996). Overview of polyethylene as abearing material: comparison of sterilization methods. Clin. Orthop. Relat. Res. *333*, 76–86.

Cotter, J.J., Maguire, P., Soberon, F., Daniels, S., O'Gara, J.P., and Cassey, E. (2011). Disinfection of meticillin-resistant Staphylococcus aureus and *Staphylococcus epidermidis* biofilms using a remote non-thermal gas plasma. J. Hosp. Infect. *78*, 204–207.

Crow, S., and Smith, J.H. 3rd. (1995). Gas plasma sterilization-application of space-age technology. Infect. Control Hosp. Epidemiol. *16*, 483–487.

De Oliveira Cardoso Macedo, M., de Macedo, H.R., Gomes, D.L., and de Freitas Daudt., Rocha, H.A., and Alves, C. Jr. (2013). Inhibition of microbial growth on chitosan membranes by plasma treatment. Art. Organs *37*, 998–1002.

Deilmann, M., Halfmann, H., Bibinov, N., Wunderlich, J., and Awakowicz, P. (2008). Low-pressure microwave plasma sterilization of polyethylene terephthalate bottles. J. Food Prot. *71*, 2119–2123.

Delgado, L.M., Pandit, A., and Zeugolis, D.I. (2014). Influence of sterilization methods on collagen-based devices stability and properties. Expert Rev. Med. Devices *11*, 305–314.

Deng, X., Shi, L., and Kong, M.G. (2006). Physical mechanisms of inactivation of Bacillus subtilis spores using cold atmospheric plasmas. IEEE Trans. Plasma Science *34*, 1310–1316.

Donohue, K.V., Bures, B.L., Bourham, M.A., and Roe, R.M. (2006). Mode of action of a novel nonchemical method of insect control: atmospheric pressure plasma discharge. J. Eco. Entomol. *99*, 38–47.

Draeger, J., and Prueter, J.W. (1990). Suitability of different procedures for the sterilization of microsurgical instruments. A review of current and future procedures [in German]. Klin. Monbl. Augenheilkd. *197*, 133–137.

Drews, R.C. (2001). Sterilization by plasma gas. Ophthalmology *108*, 1011.

Du, C., Shi, T., Sun, Y., and Zhuang, X. (2008). Decolorization of acid orange 7 solution by gas–liquid gliding arc discharge plasma. J. Hazard. Mater. *154*, 1192–1197.

Duffy, R.E., Brown, S.E., Caldwell, K.L., Lubniewinski, A., Anderson, N., Edelblute, C.M., Malik, M.A., and Heller, L.C. (2014). Surface-dependent inactivation of model microorganisms with shielded sliding

plasma discharge and applied air flow. Bioelectrochemistry. 2014 Aug 23. Pil:S1567–5394(14)00129–7. Doi;10.1016
Edelhausser, H., Holley, G., Tess, A., Divan, H., Helmy, M., Arduino, M., and Jarvis, W.R. (2000). Arch. Ophthalmol. *118*, 1167–1176.
Ehrman, J. (1989). Gas plasma microwave sterilizes instruments. Dentist *67*, 23–27.
Ermolaeva, S.A., Sysolyatina, E.V., Kolkova, N.I., Bortsov, P., Tuhvatulin, A.I., Vasillev, M.M., Mukhachev, A.Y., Petrov, O.F., Tetsyji, S., Naroditsky, B.S., Morfill, G.E., and Fortov, V.E. (2012). Non-thermal argon plasma is bacteriocidal for the intracellular bacterial pathogen Chlamydia trachomatis. J. Med. Microbiol. *61*, 793–799.
Ermolaeva, S.A., Petrov, O.F., Miller, G.G., Shaginian, I.A., Naroditskii, B.S., Sysoliatina, E.V., Mukhachev, Ala, Morfill, G.E., Fortov, V.E., Grigor'ev, A.J., and Gintsburg, A.L. (2011). Prospects for the use of low-temperature gas plasma as an antimicrobial agent. Vestn Ross. Akad. Med. Nauk. (10), 15–21. Russian.
Ermolaeva, S.A., Varfolomeev, A.F., Chernukha, M.Y., Yurov, D.S., Vasllev, M.M., Kaminskaya, A.A., Moisenovich, M.M., Romanova, J.M., Murashev, A.N., Selezneva, I.I., *et al.* (2011). Bacteriocidal effects of non-thermal argon plasma in vitro, in biofilm and in the animal model of infected wounds. J. Med. Microbiol. *60*, 75–83.
Fan, X., Zhu, T., Sun, Y., and Yan, X. (2011). The roles of various plasma species in the plasma and plasma-catalytic renival of low-concentration formaldehyde in air. J. Hazard. Mater. 2011 Nov 30;196:380–385. Doi:10.1016.
Fernandez, A., Noriega, E., and Thompson, A. (2013). Inactivation of *Salmonella enterica* serovar Typhimurium on fresh produce by cold atmospheric gas plasma technology. Food Microbiol. *33*, 24–29.
Fernandez, A., Shearer, N., Wilson, D.R., and Thompson, A. (2011). Effect of microbial loading on the efficiency of cold atmospheric gas plasma inactivation of *Salmonella enterica* serovar Typhimurium. Int. J. Food Microbiol. *152*, 175–180.
Ferreira, S.D., dernell, W.S., Powers, B.E., Schochel, R.A., Kuntz, C.A., Withrow, S.J., and Wilkins, R.M. (2001). Effect of gas plasma sterilization on the osteoinductive capacity of demineralized bone matrix. Clin. Orthop. Relat. Res. *388*, 233–239.
Fisher, J., Reeves, E.A., Issac, G.H., Saum, K.A., and Sanford, W.M. (1997). Comparison of the wear of aged and non-aged ultrahigh molecular weight polyethylene sterilized by gamma irradiation and by gas plasma. J. Mater. Sci. Mater. Med. *8*, 375–378.
Frazer, L. (1999). Plasma jet takes off. Environ Health Perspect. *107*, A414–415.
Fricke, K., Koban, I., Tresp, H., Jabionowski, L., Schroeder, K., Kramer, A., Weltmann, K.D., von Woedtke, T., and Kocher, T. (2012). Atmospheric pressure plasma: a high-performance tool for the efficient removal of biofilms. PLos One 7; e42539. Doi:10.1371
Gao, J., Yu, J., Li, Y., He, X., Bo, L., Pu, L., Yang, W., Lu, Q., and Yang, Z. (2006). Decoloration of aqueous Brilliant Green by using glow discharge electrolysis. J. Hazard Mater. *137*, 431–436.
Hoffmann, C., Berganza, C., and Zhang, J. (2013). Cold atmospheric plasma: methods of production and application in densitry and oncology. Med. Gas Res. *3*, 21.
Isbary, G., Shimizu, T., Li, Y.F., Stolz, W., Thomas, H.M., Morfill, G.E., and Zimmermann, J.L. (2013). Cold atmospheric plasma devices for medical issues. Expert. Rev. Med. Devices *10*, 367–377.
Gatineau, M., El-Warrak, A.O., Bolliger, C., Mourez, M., and Berthiaume, F. (2012). Effects of sterilization with hydrogen @eroxide gas plasma, ethylene oxide, and steam on bioadhesive properties of nylon and polyethylene lines used for stabilization of canine stifle joint. Am. J. Vet. Res. 73, 1665–1669.
Geiss, H.K. (1994). New sterilization technology-are they applicable for endoscopic surgical instruments? Endosc. Surg. Allied. Technol. *2*, 276–278.
Graham, G.S., and Mieinik, T.J. (1997). Industrial low-temperature gas plasma sterilization. Med. Device Technol. *8*, 28–30.
Griffiths, N. (1993). Low temperature sterilization using gas plasmas. Med. Devices. Technol. *4*, 37–40.
Guo, Y., Liao, X., and Te, D. (2008). Detection of hydroxy radical in plasma reaction on toluene removal. J. Environ. Sci. *20*, 1429–1432.
Han, L., Patil, S., Keener, K.M., Cullen, P.J., and Bourke, P. (2014). Bacteial inactivation by high-voltager atmospheric cold plasma; influence of process parameters and effects on cell leakage and DNA. J. Appl. Microbiol. *116*, 784–794.
Harkln, D.W., Barros D'Sa, A.A., McCalllon, K., Hoper, M., Halliday, M.I., and Campbell, F.C. (2001). Bacteriocidal/permeability-increasing protein attenuates systemic inflammation and acute lung injury in porcine lower limb ischemia–reperfusion injury. Ann. Surg. *234*, 233–244.

Hauser, J., Halfmann, H., Awakowicz, P., Koeller, M., and Esenwein, S.A. (2008). A double inductively coupled low-pressure plasma for sterilization of medical implant materials. Biomed. Technol. (Vel). *53*, 199–203.

Hauser, J., Esenwein, S.A., Awakowicz, P., Steinau, H.U., Koeller, M., and Halfmann, H. (2011). Stertilization of heat-sensitive silicone implant material by low-pressure gas plasma. Biomed. Instrument. Technol. *45*, 75–79.

Hayes, J., Kirf, D., Garvey, M., and Rowan, N. (2013). Disinfection and toxicological assessments of pulsed UV and pulsed-plasma gas-discharge treated-water containing the waterborne protozoan enteroparasite *Cryptosporidium parvum*. J. Microbiol. Methods *94*. 325–337.

Hernandez-Arias, A.N., Rodriguez-Mendez, B.G., Lopez-Callejas, R., Alcantara-Diaz, D., Valencia-Alvarado, R., Mercado-Cabrera, A., Pena-Eguiluz, R., Munoz-Castro, A.E., Barocio, S.R., and de la Piedad-Beneite (2012). Inactivation of *Escherichia coli* in water by pulsed dielectric barrier discharge in coaxial reactor. J. Water Health *10*, 371–379.

Hesby, R.M., Hagamann, C.R., and Stanford, C.M. (1997). Effects of radiofrquency glow discharge on impression material surface wettability. J. Prosthet. Dent. *77*, 414–422.

Hong, Y.F., Kang, J.G., Lee, H.Y., Uhm, H.S., Moon, E., and Park, Y.H. (2009). Sterilization effect of atmospheric plasma on E. coli and Bacilus subtilis endospores. Lett. Appl. Microbiol. *48*, 33–37.

Hopper, R.H. Jr., Young, A.M., Orishimo, K.F., and Engh, C.A. Jr. (2003). Effect of terminal sterilization with gas plasma or gamma radiation on wear of polyethylene liners. J. Bone Joint Surg. Am. *85*, 464–468.

Hosseinzadeh Colagar, A., Memariani, H., Sohbatzadeh, F., and Valinataj Omran, A. (2013). Nonthermal atmospheric argon plasma jet effects on Escherichia coli biomacromolecules. Appl. Biochem. Biotechnol. *171*, 1617–1629.

Hury, S., Vidal, D.R., Desor, F., Pelletier, J., and Lagarde, T. (1998). A parametric study pf the destruction efficiency of *Bacillus* spores in low pressure oxygen-based plasmas. Lett. Appl. Microbiol. *26*, 417–421.

Idlibi, A.N., Al-marrawi, F., Hanning, M., Lehmann, A., Rueppell, A., Schindler, A., Jentsch, H., and Rupf, S. (2013). Destruction of oral biofilms formed in situ on machined titanium (Ti) surfaces by cold atmospheric plasma. Biofouling *29*, 369–379.

Joaquin, J.C., Kwan, C., Abramzon, N., Vandervoort, K., and Brelles-Marino, G. (2009). Is gas-discharge plasma a new solution to the old problem of biofilm inactivation? Microbiology *155*, 724–732.

Jungersten, L., Edlund, A., Peterson, A.S., and Wennmaim, A. (1996). Plasma nitrate as an index of nitric oxide formation in man: analyses of kinetics and confounding factors. Clin. Physiol. *16*, 369–379.

Justan, I., Cernohorska, L., Dvorak, Z., and Slavicek, P. (2014). Plasma discharge and time-dependence of its effects to bacteria. Folia Microbiol. *59*, 315–320. Praha.

Kanemitsu, K., Imasaka, T., Ishikawa, S., Kunisshima, H., Harigae, H., Ueno, K., Takemura, H., Hirayama, Y., and Kaku, M. (2005). A comparative study of ethjylene oxide gas, hydrogen peroxide gas plasma, and low-temperature steam formaldehyde sterilization. Infect. Control. Hosp. Epidemiol. *26*, 486–489.

Kang, M.H., Hong, Y.J., Attri, P., Sim, G.B., Lee, G.J., Panngom, K., Kwon, G.C., Choi, E.H., Uhm, H.S., and Park, G. (2014). Analysis of the antimicrobial effects of nonthermal plasma on fungal spore in ionic solutions. Free Radic. Biol. Med. *72*, 191–199.

Kawamura, K., Sakuma, A., Nakamura, Y., Oguri, T., Sato, N., and Kido, N. (2012). Evaluation of bactericidal effects of low-temperature nitrogen gas plasma towards application to short-time sterilization. Microbiol. Immunol. *56*, 431–440.

Kim, J.S., Lee, E.J., and Kim, Y.J. (2014). Inactivation of Campylobacter jejuni with dielectric barrier discharge plasma using air and nitrogen gases. FoodBorne Pathog Dis. *11*, 645–651.

Kim, J.E., Lee, D.U., and Min, S.C. (2014). Microbial decontamination of red pepper powder by cold plasma. Food Microbiol. *38*, 128–136.

Kim, B., Yun, H., Jung, S., Jung, Y., Jung, H., Choe, W., and Jo, C. (2011). Effect of atmospheric pressure plasma on inactivation of pathogens inoculated onto bacon using two different gas compositions. Food Microbiol. *28*, 9–13.

Kim, C.H., Bahn, J.H., Lee, S.H., Kim, G.Y., Jun, S.I., Lee, K., and Baek, S.J. (2010). Induction of cell growth arrest by atmospheric non-thermal plasma in colorectal cancer cells. J. Biotechnol. *150*, 530–538.

Kinnari, T.J., Esreban, J., Zamora, N., Fernandez, R., Lopez-Santos, C.D., Yubero, F., Mariscal, D., Puertolas, J.A., and Gomez-Barrena, E. (2010). Effect of surface roughness and sterilization on bacterial adherence to ultra-high molecular weight polyethylene. Clin. Microbiol. Infect. *16*, 1036–1041.

Klaempfl, T.G., Isbary, G., Shimizu, T., Li, Y.-F., Zimmermann, J.L., Stolz, W., Schlegel, J., Morfill, G.E., and Schmidt, H.-U. (2012). Cold atmospheric air plasma sterilization against spores and other microorganisms of clinical interest. Appl. Environ. Microbiol. *78*, 5077–5082.

Kuo, X., and Lehotay, D.C. (1997). Determination of hydroxy radicals using salicylate as a trapping agent by gas chromatography–mass spectrometry. Clin. Biochem. *30*, 41–46.
Kvam, E., Davis, B., Mondello, F., and Garner, A.L. (2012). Nonthermal atmospheric plasma rapidly disinfects multidrug-resistant microbes by inducing cell surface damage. Antimicrob. Agent Chemother. *56*, 2028–2036.
Lackmann, J.W., and Bandow, J.E. (2014). Inactivation of microbes and macromolecules by atmospheric-pressure plasma jets. Appl. Microbiol. Biotechnol. *98*, 6205–6213.
Lassen, K.S., Nordby, B., and Gruen, R. (2003). Optimization of a RF-generated CF4/O2 gas plasma sterilization process. J. Biomed. Mater. Res. *65*, 239–244.
Lassen, K.S., Nordby, B., and Gruen, R. (2005). The dependence of the sporcidal effects on the power and pressure of RF-geberated plasma processes. J. Biomed. Mater. Res. *74*, 553–559.
Lassen, K.S., Johansen, J.E., and Gruen, R. (2006). Comparison of two radio-frequency plasma sterilization processes using microspot evaluation of microbial inactivation. J. Biomed. Mater. Res. *78*, 189–196.
Lee, H.J., Jung, H., Choe, W., Ham, J.S., Lee, J.H., and Jo, C. (2011). Inactivation of *Listeria monocytogenes* on agar and processed meat surfaces by atmospheric pressure plasma jets. Food Microbiol. *28*, 1468–1471.
Lee, K., Paek, K.H., Ju, W.T., and Lee, Y. (2006). Sterilization of bacteria, yeast and bacterial endospores by atmospheric-pressure cold plasma using helium and oxygen. J. Microbiol. *44*, 269–275.
Lee, J.K., and Chol, C.H. (2012). Two-stage reimplantation in infected total knee arthroplasty using a re-sterilized tibial polyethylene insert and femoral component. J. Arthroplasty, *27*, 1701–1706.
Lerouge, S., Fozza, A.C., Wertheimer, M.R., Marchand, R., Tabrizian, M., and Yahia, L'H. (1999). Plasma sterilization; Spore Destruction by Microwave Plasmas, MRS Proceedings, 544 in Plasma Decomposition and Treatment of Polymers, R. d'Agostino, W.W. Lee, B.D. Ratner, and M.R. Wertheimer, eds (Materials Research Society).
Lerouge, S., Wertheimer, M.R., Marchand, R., Tabrizian, M., and Yahia, L. (2000). Effect of gas composition on spore mortality and etching during low-pressure plasma sterilization. J. Biomed. Mater. Res. *51*, 128–135.
Liang, W.J., Li, J., Li, J.X., Zhu, T., and Jin, Y.Q. (2010). Formaldehyde removal from gas streams by means of $NaNO_2$ dielectric barrier discharge plasma. J. Hazard. Mater. *175*, 1090–1095.
Liao, W.T., Lee, W.J., Chen, C.Y., and Shih, M. (2001). Decontamination of ethylene oxide in the RF plasma environment. Environment. Technol. *22*, 165–173.
Liu, D., Xiong, Z., Du, T., Zhou, X., Cao, Y., and Lu, X. (2011). Bacterial-killing effect of atmospheric pressure non-equilibrium plasma jet and oral mucosa response. J. Huazhong Univ. Sci. Technol. Med. Sci. *31*, 852–856.
Lleixa Calvet, J., Grafahrend, D., Klee, D., and Moeller, M. (2008). Sterilization effect on starPEG coated polymer surfaces: characterization and cell viability. J. Mater. Sci. Mater. Med. *19*, 1631–12636.
Lupu, A.R., and Georgescu, N. (2010). Cold atmospheric plasma jet effects on V79–4 cells. Roum. Arch. Microbiol. Immunol. *69*, 67–74.
McCullough, K.D., and Bartfay, W.J. (2007). The dose-dependent effects of chronic iron overload on the production of oxygen free radicals and vitamin E concentration in the liver of a murine model. Biol. Res. Nurs. *8*, 300–304.
MacDonald, D., Hanzlik, J., Sharkey, P., Parvizi, J., and Kurtz, S.M. (2012). In vivo oxidation and surface damage in retrieved ethylene oxide-sterilized total knee arthroplasties. Clin. Orthop. Relat. Res. *470*, 1826–1833.
Magureanu, M., Piroi, D., Mandache, N.B., david, V., Medvedovici, A., Bradu, C., and Parvulescu, V.L. (2011). Degradation of antibiotics in water by non-thermal plasma treatment. Water Res. *45*, 3407–3416.
Manolache, S., Somers, E.B., Wong, A.C., Shamamian, V., and Denes, F. (2001). Dense medium plasma environments; a new approach for the disinfection of water. Environ. Sci. Technol. *35*, 3780–3785.
Markowicz, M., Koellensperger, E., Steffens, G.C., Frentz, M., Schrage, N., and Pallua, N. (2006). The impact of vacuum freeze-drying on collagen sponges after gas plasma sterilization. J. Biomater. Sci. Polym. Ed. *17*, 61–75.
Matthes, R., Huebner, N.D., Bender, C., Koban, I., Weltmann, K.D., and Kramer, A. (2010). Screening test for quality control of surface barrier discharged plasma sources with the microorganism-agar test (MAT). GMS Krankenhhyg Interdiszip, 2010 Sep 21;5(2). Pii;Doc02 doi:10.3205
Matthes, R., Bekeschus, S., Bendesr, C., Koban, I., Huebner, N.O., and Kramer, A. (2012). Pilot-study on the influence of carrier gas and plasma application (open resp delimited) modifications on physical plasma and its antimicrobial effect against *Pseudomonas aeruginosa* and *Staphylococcus aureus*. GMS Krankenhhyg. Interdiszip. 2012;7(1):Doc02.doi;10.3205

Matthes, R., Huebner, N.O., Bender, C., Koban, I., Horn, S., Bekeschus, S., Weltmann, K.D., Kocher, T., Kramer, A., and Assadian, O. (2014). Efficacy of different carrier gases for barrier discharge plasma generation compared to chlorohexidine on the survival of *Pseudomonas aeruginosa* embedded in biofilm in vitro. Skin Pharmacol. Physiol. *27*, 148–157.

Mishra, S., Viano, A., Fore, N., Lewis, G., and Ray, A. (2003). Influence of lamella features of UHMWPE on its physical and uniaxial tensile properties. I. Effect of sterilization method in uncrosslinked and unaged materials. Biomed. Mater. Eng. *13*, 135–146.

Mois, M., Mastwijk, H., Nierop Goot, M., and Abee, T. (2013). Physiological and transcriptional response of Bacillus cereus treated with low temperature nitrogen gas plasma. J. Appl. Microbiol. *115*, 689–702.

Moisan, M., barbeau, J., Moreau, S., Pelletier, J., Tabrizian, M., and Yahia, L.H. (2001). Low temperature sterilization using gas plasmas: a review of the experiments and an analysis of the inactivation mechanisms. Int. J. Pharm. *226*, 1–21.

Mols, M., Mastwijk, H., Nierop, G.M., and Abee, T. (2013). Physiological and transcriptional response of Bacillus cereus treated with low-temperature nitrogen gas plasma. J. Appl. Microbiol. *115*, 689–702.

Muranyi, P., Wunderlich, J., and Langowski, H.C. (2010). Modification of bacterial structures by a low temperature gas plasma and influence on packing material. J. Appl. Microbiol. *109*, 1875–1885.

Muranyi, P., Wunderlich, J., and Haise, M. (2008). Influence of relative gas humidity on the inactivation efficiency of a low temperature gas plasma. J. Appl. Microbiol. *104*, 1659–1666.

Muranyi, P., Wunderlich, J., and Heise, M. (2007). Sterilization efficiency of a cascaded dielectric barrier discharge. J. Appl. Microbiol. *103*, 1535–1544.

[No author listed] (1999). Choosing a low-temperature sterilization technology. Health Devices *28*, 430–455.

Niemira, B.A., Boyd, G., and Sites, J. (2014). Cold plasma rapid decontamination of food contact surfaces contaminated with Salmonella biofilms. J. Food Sci. *79*, M917–922.

Niemira, B.A. (2012). Cold plasma reduction of Salmonella and Escherichia coli 0157:H7 on almonds using ambient pressure gases. J. Food Sci. *77*, M171–175.

Niemira, B.A. (2011). Cold plasma decontamination of foods. Ann. Rev. Food Technol. *3*, 125–142.

Niemira, B.A. (2012). Cold plasma decontamination of foods. Ann. Rev. Food Sci. Technol. *3*, 125–142.

Nishioka, T., Takai, Y., Kawaradani, M., Okada, K., Tanimoto, H., Misawa, T., and Kusakari, S. (2014). Seed disinfection effect of atmospheric pressure plasma and low pressure plasma on Rhizoctonia soiani. Biocontrol Science *19*, 99–102.

Noriega, E., Shama, G., Laca, A., Diaz, M., and Kong, M.C. (2011). Cold atmospheric gas plasma disinfection of chicken meat and chicken skin contaminated with *Listeria innocua*. Food Microbiol. *28*, 1293–1300.

Patil, S., Moiseev, T., Misra, N.N., Cullen, P.J., Mosnier, L.P., Keener, K.M., and Bourke, P. (2014). Influence of high voltage atmospheric cold plasma process parameters and role of relative humidity on inactivation of *Bacillus atrophaeus* spore inside a sealed package. J. Hosp. Infect. Pii:S0195-6701(14)00271-0.

Peniston, S.J., and Chpi, S.J. (2007). Effect of sterilization on the physicochemical properties of molded poly(L-lactic acid). J. Biomed. Mater. Res. *80*, 67–77.

Perni, S., Shama, G., and Kong, M.G. (2008). Cold atmospheric plasma disinfection of cut fruit surfaces contaminated with migrating microorganisms. J. Food Prot. *71*, 1619–1625.

Pignata, C., D'Angelo, D., Basso, D., Cavallero, M.C., Beneventi, S., Tartaro, D., Meineri, V., and Gilli, G. (2014). Low-temperature, low-pressure gas plasma application on Aspergillus brasiliensis, Escherichia coli and pistachios. J. Appl. Microbiol. *116*, 1137–1148.

Pokorny, D., Slouf, M., and Fulin, P. (2012). Current knowledge on the effect of technology and sterilization on the structure, properties and longevity of UHMWPE in total joint replacement. Acta Chir Orthop. Traumatol. Cech. *79*, 213–221.

Ortoneda, M., O'Keeffe, S., Cullen, J.D., Al-Shamma'a A.L., and Phipps, D.A. (2008). Experimental investigations of microwave plasma UV lamp for food applications. J. Microw. Power Electromagn. Energy *42*, 13–23.

Purevdorj, D., Igura, N., Ariyada, O., and Hayakawa, I. (2003). Effect of feed gas composition of gas discharge plasmas on *Bacillus pumilus* spore mortality. Lett. Appl. Microbiol. *37*, 31–34.

Rainer, A., Centola, M., Spadaccio, C., Gherardi, G., Genovese, J.A., Licoccia, S., and Trombetta, M. (2010). Comparative study of different technique for sterilization of poly-L-lactide electrospun microfibers: effectiveness vs. material degradation. Int. J. Art. Organs 33, 76–85.

Rederstorff, E., Fatimi, A., Sinquin, C., Ratiskol, J., Merceron, C., Vinatier, C., Weiss, P., and Collec-Jouault, S. (2011). Sterilization of exopolysaccarides produced by deeep-sea bacteria: impact on their stability and degradation. Mar. Drugs *9*, 224–241.

Reeves, E.A., Harton, D.C., FizPatrick, D.P., and Fisher, J. (2000). Comparison of gas plasma and gamma irradiation in air sterilization on the delamination wear or the ultra-high molecular weight polyethylene used in knee replacements. Proc. Inst. Mech. Eng. H. *214*, 249–255.

Rossi, F., Kylian, O., and Hasiwa, M. (2008). Mechanism of Sterilization and Decontamination of Surfaces by Low-Pressure Plasma, in Advanced Plasma Technology, R. d'Agostino, P. Favia, Y. Kawai, H. Ikegami, N. Sato, and F. Arefi-Khonsari, eds (Wiley-VCH Verlag GmbH & Co. KGaA), Weinheim, Germany, pp. 319–340.

Rossi, R., and Kylian, O. (2012). Sterilization and decontamination of surfaces by plasma discharges. In Sterilization of Biomaterials and Medical Devices. Lerouge, S., and Simmons, A. eds. (Woodhead Publishers), Oxford, UK, pp. 130–131.

Roth, S., Feichtinger, J., and Hertel, C. (2010). Characterization of *Bacillus subtilis* spore inactivation in low-pressure gas plasma sterilization processed. J. Appl. Microbiol. *108*, 521–531.

Rowan, N.J., Espie, S., Harrower, J., Farrell, H., Marsili, J., Anderson, J.G., and MacGregor, S.J. (2008). Evidence of lethal and sublethal injury in food-borne bacterial pathogens exposed to high-intensity pulsed-plasma gas discharges. Lett. Appl. Microbiol. *46*, 80–89.

Rowan, N.J., Espie, S., Harrower, J., Anderson, J.G., Marsill, L., and MacGregor, S.J. (2007). Pulsed-plasma gas-discharge inactivation of microbial pathogens in chilled poultry wash water. J. Food Prot. *70*, 2805–2810.

Rubbo, H., Parthasarathy, S., Barnes, S., Kirk, M., Kalyanaraman, B., and Freeman, B.A. (1995). Nitric oxide inhibition of lipoxygenase-dependent liposome and low-density lipoprotein oxidation; termination of radical chain propagation reactions and formation of nitrogen-containing oxidized lipid derivaties. Arch. Biochem. Biophys. *324*, 15–25.

Sakudo, A., Higa, M., Maeda, K., Shimizu, N., Imanishi, Y., and Shintani, H. (2013). Sterilization mechanism of nitrogen gas plasma: induction of secondary structural change in protein. Microbiol. Immunol. *57*, 536–542.

Schultz-Jensen, N., Leipold, F., Bindslev, H., and Thomsen, A.B. (2011). Plasma-assisted pretreatment of wheat straw. Appl. Biochem. Biotechnol. *163*, 558–572.

Selcuk, M., Oksuz, L., and Basaran, P. (2008). Decontamination of grains and legumes with *Aspergillus* spp. and *Penicillium* spp. by cold plasma treatment. Bioresour. Technol. *99*, 5104–5109.

Shen, Y., Lei, L., Zhang, X., Zhou, M., and Zhang, Y. (2008). Effect of various gases and chemical catalysts on phenol degradation pathways by pulsed electrical discharges. J. Hazard. Mater. *150*, 713–722.

Shi, X.M., Zhang, G.J., Yuan, Y.K., Ma, Y., Xu, G.M., and Gu, N. (2009). Inactivation of bacterial spores using low-temperature plasma [in Chinese]. Nan Fang Yi Ke Da Xue Xue Bao. *29*, 2033–2036.

Shimizu, K., Yano, H., Nakamura, E., and Kaku, N. (2001). Lipid extracted freeze-dried bank bone sterilized with low temperature plasma. Ann. Transport. *6*, 26–32.

Shintani, H., and McDonnell, G.E. (2011). Definition of Sterilization, Disinfection, Decontamination, and Antisepsis In Sterilization and Disinfection by Plasma, Sterilization Mechanisms, Biological and Medical Applications, A. Sakudo and H. Shintani, eds (NOVA Science Publishers), New York.

Shintani, H., Sakudo, A., Burke, P., and McDonnell, G.E. (2010). Gas plasma sterilization of microorganisms and mechanims of action. Experiment. Therapeut. Med.*1*, 731–738.

Shintani, H., Shimizu, N., Imanishi, Y., Sekiya, T., Tamazawa, K., Taniguchi, A., and Kido, A. (2007). Inactivation of microorganisms and endotoxins by low temperature nitrogen gas plasma exposure. Biocontrol Sci. *12*, 131–143.

Shintani, H., Kurosu, S., Miki, A., Hayashi, F., and Kato, S. (2006). Sterilization efficiency of the photocatalyst against environmental microorganisms in a health care facility. Biocontrol Science, *11*, 17–26.

Shintani, H., Taniai, E., Miki, A., Kurosu, S., and Hayashi, F. (2004). Comparison of the collecting efficiency of microbiological air samplers, J. Hosp. Infect. *56*, 42–48.

Shintani, H., and Akers, J.E. (2000). On the cause of performance variation of biological indicator used for sterility assurance. PDA J Pharm. Sci. Technol. *54*, 332–342.

Shintani, H., Sasaki, K., Kajiwara, Y., Itoh, J., Takahashi, M., and Kokubo, M. (2000). Validation of D value by different SCD culture medium manufacturer and/or different SCD culture medium constitute. PFA J. Pharm. Sci. Technol. *54*, 6–12.

Shintani, H. (1997). The use of two or more microorganisms versus one microorganism in the carrier materials for biological indicators. Biomed. Instrument. Technol. *31*, 387–390.

Shintani, H. (1996). Factors in the preparation of biological indicators that affect the decimal reduction time. Biomed. Instrument. Technol. *30*, 449–453.

Silva, J.M., Moreira, A.J., Oliveira, D.C., Bonato, C.B., Mansano, R.D., and Pinto, T.J. (2007). Comparative sterilization effectiveness of plasma in O_2–H_2O_2 mixtures and ethylene oxide treatment. PDA J. Pharm. Technol. *61*, 204–210.

Smith, C.A., Khoury, J.M., Shields, S.M., Roper, G.J., Duffy, R.E., Edelhauser, H.F., and Lubniewski, A.J. (2000). Unexpected cornel endothelial cell decompensation after intraocular surgery with instruments sterilized by plasma gas. Opthalmology *107*, 1561–1566.

Southwood, L.L., and Baxter, G.M. (1996). Instrument sterilization, skin preparation, and wound management. Vet. Clin. North Am. Equine Pract. *12*, 173–194.

Stapelmann, K., Fiebrandt, M., Raguse, M., Awakowicz, P., Reitz, G., and Moeller, O. (2013). Utilization of low-pressure plasma to inactivate bacterial spores on stainless steel screws. Astrobiology, *13*, 597–606.

Tank, C., Raman, S., Karan, S., Gosavi, S., Lalla, N.P., Sathe, V., Berndt, R., Gade, W.N., Bhoraskar, S.V., and Mathe, V.L. (2013). Antimicrobial activity of silica coated silicon nano-tubes (SCSNT) and silica coated silicone nano-particles (SCSNP) synthesized by gas phase condensation. J. Mater. Sci. Mater. Med. *24*, 1483–1490.

Tseng, S., Abramzon, N., Jackson, J.O., and Lin, W.J. (2012). Gas discharge plasmas are effective in inactivating Bacillus and Clostridium spores. Appl. Microbiol. Biotechnol. *93*, 2563–2570.

Stoffels, E., Kieft, I.E., Siadek, R.E., van der Laan, E.P., and Slaaf, D.W. (2004). Gas plasma treatment: a new approach to surgery? Crit. Rev. Biomed. Eng. *32*, 427–460.

Sun, K., Yang, X., Ye, G., Pan, H., and Wang, J. (2013). Evaluation of two different cold plasma treatments on root canal infection with *Enterococcus faecalis* biofilms [in Chinese]. Hua Xi Kou Qiang Yi Xue Za Zhi. *31*, 195–198.

Tessarolo, F., Caola, I., Caciagli, P., Guarrera, G.M., and Nollo, G. (2006). Sterility and microbiological assessment of reused single-use cardiac electrophysiology catheters. Infect. Control Hosp. Epidemiol. *27*, 1385–1392.

Thiyagarajan, M., Sarani, A., and Gonzales, X.F. (2013). Characterization of an atmospheric pressure plasma jet and its applications for disinfection and cancer treatment. Stud. Health Technol. Inform. *184*, 443–449.

Traba, C., Chen, L., and Liang, J.F. (2013). Low power gas discharge plasma mediated inactivation and removal of biofilms formed on biomaterials. Curr. Appl. Phys. *13* (Suppl. 1), S12–S18.

Trostle, S.S., Hendrickson, D.A., and Franke, C. (2002). The effects of ethylene oxide gas and gas plasma sterilization on failure strength and failure mode of pre-tied monofilament ligature loops. Vet. Surg. *31*, 281–284.

Vandamme, M., Robert, E., Lerondel, S., sarron, V., Ries, D., Dozias, S., Soblio, J., Gosset, D., Kieda, C., Legrain, B., Pouvesie, J.M., and Pape, A.L. (2012). ROS implantation in a new antitumor strategy based on non-thermal plasma. Int. J. Cancer *130*, 2185–2194.

Vandervoort, K.G., and Brelles-Marino, G. (2014). Plasma-mediated inactivation of *Pseudomonas aeruginosa* biofilms grown on borosilicate surfaces under continuous culture system. PLoS One 2014 Oct 10 9(10);e108512.doi;10.1371

Vetten, M.A., Yah, C.S., Singh, T., and Gulumian, M. (2014). Challenges facing sterilization and depyrogenation of nanoparticles: Effects on structural stability and biomedical applications, Nanomedicine *1000*, 1391–1399.

Wang, S., Wang, W.C., Yang, D.Z., Liu, Z.J., and Zhang, S. (2014). Optical and application study gas–liquid discharge excited by bipolar nanosecond pilse in atmospheric air. Spectrochim. Acta A Mol Biomol. Spectrosc. *131*, 571–576.

Wang, C.H., Li, G.F., We, Y., and Wang, Y. (2008). Growth inhibition of Microcystis aeruginosa in packed-bed discharge plasma reactor. Huan Jing Ke Xue., *29*, 368–374. Chinese.

Whittaker, A.G., Graham, E.M., Baxter, R.L., Jones, A.C., Richardson, P.R., Merek, G., Campbell, G.A., Alken, A., and Baxter, H.C. (2004). Plasma cleaning of dental instruments. J. Hosp. Infect. *56*, 37–41.

Wiegand, C., Beier, O., Horn, K., Pfuch, A., Toelke, T., Hipier, U.C., and Schimanski, A. (2014). Antimicrobial impact cold atmospheric plasma on medical critical yeasts and bacteria cultures. Skin Pharmacol. Physiol. *27*, 25–35.

Winter, T., Bernhardt, J., Winter, J., Maeder, U., Schlueter, R., Weltmann, K.D., Hecker, M., and Kusch, H. (2013). Common versus noble *Bacillus subtilis* differently responds to air and argon gas plasma. Proteomics *13*, 2608–2621.

Winter, T., Winter, J., Polak, M., Kusch, K., Maeder, U., Sietmann, R., Ehibeck, J., van Hijum, S., Weltmann, K.D., Hecker, M., and Kusch, H. (2011). Characterization of the global impact of low temperature gas plasma on vegetative microorganisms. Proteomics *11*, 3518–3530.

Wittenburg, G., Lauer, G., Oswald, S., Labudde, D., and Franz, C.M. (2014). Nanoscale topographic changes on sterilized glass surfaces affect cell adhesion and spreaing. J. Biomed Mater. Res. *102*, 2755–2766.

Yamamoto, T. (1999). Optimization of non-thermal plasma for the treatment of gas streams. J. Hazard Mater. *67*, 165–181.

Yang, S.X.L., Bal, M.D., and Han, F. (2010). Rapid elimination of surface organophosphorous pollutants using hydroxy radical. Huan Jing Ke Xue. *31*, 1670–1674. Chinese.

Yet-Pole, I., Liu, Y.C., Han, K.Y., and She, T.C. (2004). Construction of a low-pressure microwave plasma reactor and its application in the treatment of volatile organic compounds. Environ. Sci. Technol. *38*, 3785–3791.

Yong, H.I., Kim, H.J., Park, S., Choe, W., Oh, M.W., and Jo, C. (2014). Evaluation of the treatment of both sides of raw chicken breasts with an atmospheric pressure plasma jet for the inactivation of *Escherichia coli*. Foodborne Pathog Dis. *11*, 652–657.

Yu, Q.S., Huang, C., Hsieh, F.H., Huff, H., and Duan, Y. (2007). Bacterial inactivation using a low-temperature atmospheric plasma brush sustained with argon gas. J. Biomed. Mater. Res. *80*, 211–219.

Yu, H., Perni, S., Shim, J.J., Wang, D.Z., Kong, M.G., and Shama, B. (2006). Effects of cell surface loading and phase of growth in cold atmospheric gas plasma inactivation of *Escherichia coli* K12. J. Appl. Microbiol. *10*, 1323–1330.

Yuan, M.H., Narengerile, Watanabe, T., and Chang, C.Y. (2010). DC water plasma at atmospheric pressure for the treatment of aqueous phenol. Environment. Sci. Technol. *44*, 4710–4715.

Yuen, P.K., Su, H., Gorai, V.N., and Fink, K.A. (2011). Three-dimentional interconnected microporous poly(dimethylsiloxane) microfluidic devices. Lab. Chip. *11*, 1541–1544.

Zhang, H., Yang, L., Yu, Z., and Huang, Q. (2014). Inactivation of *Microcystis aeruginosa* by DC glow discharge plasma; impacts on cell integrity, pigment contents and microcystins degradation. J. Hazard Mater. 2014 Mar 15:268:33-42. Doi: 10.1016

Zhang, X., Zhang, X.F., Li, H.P., Wang, L.Y., Zhang, C., Xing, X.H., and Bao, C.Y. (2014). Atmospheric and room temperature plasma (ARTP) as a new powerful mutagenesis tool. Appl. Microbiol. Biotechnol. *98*, 5387–5396.

Zhang, R.B., Wu, Y., Li, G.F., Wang, N.H., and Li, J. (2004). Plasma induced degradation of indigo Carmine by bipolar pulsed dielectric barrier discharge (DBD) in the water–air mixture. J. Environ. Sci. (China). *16*, 808–812.

Zhao, J., and Lubman, D.M. (1993). Detection of liquid injection using an atmospheric pressure ionization radiofrequency plasma source. Anal. Chem. *65*, 866–876.

Current Progress in Advanced Research into the Inactivation of Fungi and Yeasts by Gas Plasma

9

Gyungsoon Park

Abstract

Recently, the risk level of fungi has been increased in human health, food safety, agriculture, and ecosystem. Efficient, eco-friendly, and long-lasting control tools for fungal diseases and contaminants are needed more than ever, and plasma has been explored as a candidate tool satisfying these criteria. In this chapter, studies on antifungal activity of plasma are summarized. Numerous studies have demonstrated that plasma treatment can efficiently inactivate fungal spores and disinfect human tissues, paper, fabrics, crop seeds, plant leaves, and foods. Plasma generated reactive species as possible fungicidal factors, may destroy or degenerate fungal cell wall and consecutively damage cell membrane and internal components. However, more experimental studies, particularly *in vivo*, examining antifungal effects of plasma are still required for proving the potentiality of plasma in fungal disease control. In addition, mechanisms of fungal sterilization by plasma should be elucidated in order to produce useful information applicable to optimization of plasma sterilization technology.

Introduction

Many yeast like and filamentous fungi cause serious diseases in human, animals and plants, and contaminate air, materials and foods. Fungal diseases have been relatively infrequent in human compared to bacterial diseases but their outcomes are much more devastating. A recent review about emerging fungal threats demonstrates that disastrous fungal diseases in nature and managed landscapes have increased in number during past two decades, and recently caused extinction of some wild species and imperilled food security (Fisher *et al.*, 2012). The level of fungal threats may be increasing in modern society as a result of human activity modifying natural environments, which provokes emergence of new fungal strains (Fisher *et al.*, 2012). In addition, there are many opportunistic fungal pathogens that are able to develop virulent infectious diseases depending on environmental changes. For example, the fungi *Pseudallescheria boydii* and *Scedosporium prolificans* became capable of infecting the pulmonary and central nervous systems in near-drowning people during the Asian and Japanese earthquakes and tsunamis (Nakamura *et al.*, 2011). After Hurricanes Katrina and Rita in New Orleans, *Trichoderma*, a benign fungus, together with *Aspergillus, Penicillium, Paecilomyces* species were widespread in flooded home, producing spore and endotoxins that were harmful to human health (Chew *et al.*, 2006; Rao *et al.*, 2007). Many weak and non-pathogenic fungi have been recently emerged as major infectious agents for plant fungal disease epidemics as shown in destruction of wood population by *Phytophthora ramorum* in California and the UK (Grunwald *et al.*, 2008).

Control tools for fungal diseases and contaminants such as antifungal drugs, fungicides, heat treatment, irradiation, and biocontrol have been developed. However, several problems related to efficiency, safety and resistance are also appeared following frequent use of these control tools. Recently, many researchers have focused on multidisciplinary approach in the development of control strategies for fungal diseases and contaminants. Plasma technology developed on the multi-disciplinary basis can be a potential control method for fungal diseases and contaminants. Biological impact of plasma has been well demonstrated in the area of medicine and agriculture, and antimicrobial effects of plasma in particular are a subject of active research (Fridman *et al.*, 2008; Ito *et al.*, 2007). In this chapter, research performed during last decade about inactivation of yeast like and filamentous fungi by plasma has been summarized and future research directions are discussed.

Fungal pathogens and human life

Fungal pathogens have a great impact on human life as agents causing serious diseases of animals, plants, and humans (Thornton and Wills, 2015). Fungal pathogens threaten human and animal health and cause significant economic losses in agriculture. Differently from bacteria (prokaryotes), fungi are included in eukaryotes and many are multicellular. There are yeast type and filamentous fungi. Many yeast type fungi are opportunistic pathogens to human and animals, and majority of plant diseases is caused by filamentous fungi.

Fungal pathogens are well known to cause serious diseases in immunocompromised humans and animal live stocks (Wilkinson, 1988). According to recent statistics, the number of life-threatening infections by opportunistic invasive and endemic dimorphic fungi exceeds two million cases per year (Brown *et al.*, 2012). Although there are dimorphic fungi such as *Blastomyces, Histoplasma, Coccidioides* and *Paracoccidioides* that can infect healthy individuals, most diseases caused by human fungal pathogens are predominantly opportunistic, attacking a weakened immune system (Pfaller and Diekema, 2005). *Candida* and *Aspergillus* species are well-known opportunistic fungal pathogens (Bennett, 2010; Odds, 1987). These species cause skin-related and respiratory diseases, and their impacts are various ranging from chronic infection to fatal diseases. Recently, fungal species causing diseases in both healthy and immunocompromised humans and animals such as *Pseudallescheria boydii* and *Scedosporium prolificans* have been emerged (Cortez *et al.*, 2008). In addition, host ranges of some fungal pathogens become broadened. *Aspergillus sydowii* causing aspergillosis can also infect the sea fan coral, and *F. solani* and *Scedosporium* species can attack other wild and domestic animals (Rypien *et al.*, 2008).

Mycotoxins produced by fungi can be an another serious health-threatening factor to both human and animals (Bennett and Klich, 2003). Species like *Alternaria, Aspergillus, Cladosporium, Fusarium, Penicillium*, and *Stachybotrys* are well-known producers of mycotoxins and allergens (Hedayati *et al.*, 2007). These mycotoxins and allergens can disturb respiratory tracts and immune responses, and contaminate foods (Hedayati *et al.*, 2007).

Plant fungal pathogens have been known as a major threat to food security and agriculture. Pathogens destroy about al least 125 million tones of major food crops (rice, wheat, maize, potato, soybean) every year and contribution by fungal pathogens may be significant

(Fisher *et al.*, 2012). The economic loss caused by fungal infections in the global production of rice, wheat, and maize is about $60 billion per year (Fisher *et al.*, 2012). The rice blast fungus, *Magnaporthe grisea*, causes 10–35% reduction in rice crops annually, and the emerging virulent strain of wheat stem rust fungus, *Puccinia graminis* Ug99, has brought almost 100% loss in crop production (Singh *et al.*, 2011; Wilson and Talbot, 2009). Many phytopathogenic fungi infect plants through specialized infection structures, and by secreting virulence factors and effectors that modulate host defences (Wilson and Talbot, 2009).

Control strategies and methodologies for fungal diseases have mainly focused on chemical based approach. Although biocontrol and irradiation based control tools are recently developed, majority of fungal prevention and eradication in practice has pretty much depended on use of antifungal chemicals and fungicides. Various antifungal agents for human fungal pathogens are developed. Amphotericin B, echinocandins (anidulafungin, caspofungin, micafungin), and azoles (itraconazole, posaconazole, voriconazole) have been actively used in the clinical treatments as antifungal therapy (Donnelly, 2013). However, efficiency is dramatically decreased for invasive fungal infections. For control of plant fungal diseases, fungicides have been frequently used. Fungicides are classified as specific or multi-sites based on target(s), and half of total fungicides used in practice are included in sterol demethylation inhibitors (DMIs) and quinone outside inhibitors (QoIs) (Hirooka and Ishii, 2013). Worldwide consumption of fungicides has been increasing by 6.5% annually since 1999 (Hirooka and Ishii, 2013).

In spite of frequent utilization, antifungal agents provoke several problems such as drug resistance and inactivation efficiency. After azole resistant *Candida* species has emerged, echinocandin is applied to eradicate azole resistant strains but recently level of resistance to both azole and echinocandin is elevated (Shor and Perlin, 2015). Fungicide resistance in phytopathogenic fungi has been continuously increased since the early 1970s and threatened the crop yields as well as quality (Ishii, 2006). For example, QoI fungicides (strobilurin fungicides), a best selling fungicide in the world, inhibit fungal respiration by attacking the Qo site of cytochrome bc1 enzyme complex. Since site-specific inhibitors can easily induce resistance in fungal pathogens, fungal resistance to QoI has emerged worldwide (Ishii, 2006). MBI-D fungicides in the control of rice blast fungus *Magnaporthe grisea* have no sign for resistance for over 30 years but a recent emergence of resistant strain in the field has caused decrease in fungicide efficacy (Sawada *et al.*, 2004).

Inactivation of fungal pathogens that are important to human health by plasma

Since chemical based control of fungal diseases has several problems, alternative methods and technologies have been searched and developed. Atmospheric pressure non-thermal plasma is known to inactivate microorganisms, and its application to eradicating microbial infections and contaminations in medicine, agriculture, and food industry has been actively explored. Sterilization using plasma has been mostly focused on bacterial inactivation and fungal inactivation by plasma is relatively less investigated.

Application of plasma technology in the control of human health-related fungi has been focused on inactivation of fungal spores that may be present in indoor air, in tissues, and on materials (Table 9.1). Many studies have demonstrated antifungal activity of plasma using

Table 9.1 Application of plasma in the control of fungal infections and contaminants related to human health

Association	Fungal species	Plasma	References
Yeast cells or Spores	*C. albicans*	Glow discharge plasma	Ohkawa *et al.* (2005, 2006)
		Corona discharge plasma	Korachi *et al.* (2009)
		Plasma jet	Xiong *et al.* (2010)
		Plasma jet	Daeschlein *et al.* (2011)
		Helium ionized plasma	Brun *et al.* (2012)
		Plasma jet	Sedghizadeh *et al.* (2012)
		High-voltage pulse plasma	Song *et al.* (2013)
	C. lypolitica	Plasma jet	Akishev *et al.* (2008)
	A. niger	Plasma jet	Akishev *et al.* (2008)
		Corona discharge plasma	Korachi *et al.* (2009)
		Microwave microplasma	Mizeraczyk *et al.* (2013), Czylkowski *et al.* (2013)
	A. fumigatus	Helium ionized plasma	Brun *et al.* (2012)
Indoor air	*A. niger*	Microwave plasma	B.J. Park *et al.* (2003), J.C. Park *et al.* (2004)
		Glow discharge plasma	Ohkawa *et al.* (2006)
		Plasma generating atomic hydrogen	Nojima *et al.* (2007)
	P. citrium	Microwave plasma	B.J. Park *et al.* (2003), J.C. Park *et al.* (2004)
		Plasma generating atomic hydrogen	Nojima *et al.* (2007)
	C. cladosporiodes	Microwave plasma	J.C. Park *et al.* (2004)
	Chaetomium sp.	Microwave plasma	J.C. Park *et al.* (2004)
Silk fabrics	*A. niger*	Microwave plasma	D.J. Park *et al.* (2008)
	P. citrinum	Microwave plasma	D.J. Park *et al.* (2008)
Paper	*A. niger*	DBD plasma	Vrajova *et al.* (2008, 2009)
Skin	*T. interdigitale*	Plasma jet	Daeschlein *et al.* (2011)
	T. rubrun	Plasma jet	Daeschlein *et al.* (2011)
	M. canis	Plasma jet	Daeschlein *et al.* (2011)
Sealed package	*C. albicans*	Strip type plasma in package	Song *et al.* (2012)
Mycotoxins		Microwave plasma	B.J. Park *et al.* (2007)

yeast type and filamentous fungi such as *Candida albicans, Candida lipolytica, Aspergillus niger, Penicillium citrinum, Aspergillus fumigatus, Cladosporium cladosporiodes, Chaetomium* sp., etc. (Table 9.1). Among these fungi, *C. albicans* has been most frequently used for inactivation by plasma. *C. albicans* is an yeast like fungus and causes opportunistic oral and genital infections as well as infections of nails and skin called candidiasis (Odds, 1987b). Experimental results demonstrate that glow discharge using RF power, corona discharge using AC/DC high-voltage power, and plasma jet can efficiently inactivate *C. albicans* cells within several minutes (Korachi *et al.*, 2009; Ohkawa *et al.*, 2005, 2006; Sedghizadeh *et al.*, 2012; Xiong *et al.*, 2010). Similar inactivation efficiency of plasma is also observed in controlling resistant *C. albicans* present in a sealed package and associated with glass plate (Song *et al.*, 2012,2013). A recent study shows that plasma significantly reduces the viability of *C. albicans* and *A. fumigatus* spores but human ocular cells and *ex vivo* corneas are not damaged by treatment with same plasma dose (Brun *et al.*, 2012). This observation suggests a possibility of clinical application in which plasma can be used as an efficient tool to disinfect ocular tissues.

Spores of fungi frequently discovered in indoor air, *A. niger* and *P. citrinum*, are considered as a potential factor threatening human health. Plasma treatment is shown to dramatically eradicate spores of *A. niger* and *P. citrinum* (Nojima *et al.*, 2007; Ohkawa *et al.*, 2006; Park *et al.*, 2003, 2004). Mycotoxins produced by fungi cause many health problems in human and animals. Inactivation of mycotoxin by plasma has been demonstrated in a study in which aflatoxin B1 (AFB1), deoxynivalenol (DON) and nivalenol (NIV) are completely removed after 5 s microwave plasma treatment and cytotoxicity of plasma-treated mycotoxin was significantly reduced (Park *et al.*, 2007).

Selective inactivation of fungi by plasma without damage on associated tissues or materials has been also observed. Clinical isolates of fungal species, *Trichophyton interdigitale, Trichophyton rubrum, Microsporum canis, C. albicans,* involved in dermatomycosis, were poorly grown after plasma treatment and *C. albicans* exhibited the largest inactivation (Daeschlein *et al.*, 2011). In addition, plasma treatment completely destroys reproductive fungal structures of *T. interdigitale* in dandruff of patients with tinea pedis (Daeschlein *et al.*, 2011). This study suggests that plasma can be a suitable antifungal tool applicable to *in vivo* treatment of skin infection. In several studies, plasma is found to efficiently remove fungal spores contaminating papers and fabrics without damaging materials (Park *et al.*, 2003, 2004, 2008; Vrajova *et al.*, 2008). This is a quite promising result for plasma to become more efficient and eco-friendly sterilization tool than traditional sanitation methodology.

Although fungal pathogens and mycotoxins are efficiently inactivated by atmospheric pressure plasma, mechanisms of plasma sterilization are still under active investigation. Microscopic data from several studies indicate that both yeast (*C. albicans*) cells and filamentous fungal spores after plasma treatment are crushed (Korachi *et al.*, 2009; Ohkawa *et al.*, 2006; Park *et al.*, 2003, 2008; Xiong *et al.*, 2010). However, cell burst and tearing are more often observed in yeast cells while shrinkage and wrinkling in filamentous fungal spores (Korachi *et al.*, 2009; Ohkawa *et al.*, 2006; Park *et al.*, 2003, 2008; Xiong *et al.*, 2010). These results indicate that fungal cell wall and membrane are damaged and intracellular components may also be affected. Morphological alteration and destruction of fungal cells may be caused by plasma produced reactive oxygen and nitrogen species (ROS and RNS), and ROS such as O radical and OH radical seem to play more critical role in antifungal activity (Ohkawa *et al.*, 2006; Song *et al.*, 2012; Xiong *et al.*, 2010). Besides reactive species, UV emitted from plasma and plasma etching are also suggested as contributors to fungal inactivation (Park *et al.*, 2003, 2004). A recent study has demonstrated that intracellular ROS level elevated by plasma treatment is a critical element for plasma sterilization (Brun *et al.*, 2012). Atomic hydrogen released by a novel plasma device can also reduce very effectively fungal contaminants (Nojima *et al.*, 2007).

Inactivation of fungal pathogens that are important in agriculture and foods by plasma

Compared to medically important fungal pathogens, crop and food associated fungal pathogens have been less explored for their control by plasma. Plasma application in the control of agricultural fungal contamination has been focused on inactivation of fungal spores *in vitro* and disinfection of seeds, crops, and foods associated with fungi (Table 9.2). Spores of *Fusarium graminearum, Fusarium oxysporum, Penicillium digitatum,* and *Cladosporium fulvum* which are major fungal pathogens of wheat, tomato, and citrus have shown to be effectively inactivated by atmospheric pressure non-thermal plasma in laboratory culture condition (Iseki *et al.*, 2010; Lu *et al.*, 2014; Na *et al.*, 2013). Spore inactivation efficiency is highly dependent on plasma density, power, frequency, treatment time and fungal species (Lu *et al.*, 2014; Na *et al.*, 2013). Particularly, plasma jet with a relatively high plasma density enable to completely kill resistant *C. fulvum* spores, and this suggests that plasma can potentially control resistant fungal strains (Lu *et al.*, 2014). In a recent study using *Ascochyta pinodella* and *Fusarium culmorum,* air plasma treatment is shown to be able to inhibit mycelial growth and alter morphology of mycelia (Avramidis *et al.*, 2010). An *in vitro* study using fungal spores and host plant has demonstrated that plasma dose for killing or inactivating fungal spores may be able to induce resistance in host plant (Panngom *et al.*, 2014).

As demonstrated in *in vitro* studies, plasma may be able to become a very efficient tool for controlling fungal diseases in agriculture and food industry. Fungal contamination on agricultural products can be a main cause for decreasing crop yield and threatening food

Table 9.2 Application of plasma in the control of fungal infections and contaminants related to agriculture and foods

Association	Fungal species	Plasma	References
Spores	*P. digitatum*	Non-equilibrium plasma jet	Iseki *et al.* (2010)
	F. graminearum	Microwave plasma	Na *et al.* (2013)
	F. oxysporum	Microwave plasma	Na *et al.* (2013)
	F. oxysporum	DBD microplasma	Panngom *et al.* (2014)
Mycelia	*A. pinodella*	DBD plasma	Avramidis *et al.* (2010)
	F. culmorum	DBD plasma	Avramidis *et al.* (2010)
Nuts	*A. parasiticus*	Low-pressure plasma	Basaran *et al.* (2008)
Seeds	*Aspergillus* spp.	Low-pressure plasma	Selcuk *et al.* (2008)
	Penicillium spp.	Low-pressure plasma	Selcuk *et al.* (2008)
	C. fulvum	Plasma jet	Lu *et al.* (2014)
	F. fujikuroi	DBD plasma	Jo *et al.* (2014)
	R. solani	Atmospheric pressure plasma, low-pressure plasma	Nishioka *et al.* (2014)
Wood	*A. pullulans*	Afterglows plasma	Lecoq *et al.* (2013)
Red pepper powder	*A. flavus*	Microwave plasma	Kim *et al.* (2014)
Infected leaf		Plasma jet	Zhang *et al.* (2014)

safety. Plant seeds and fruits are very vulnerable to fungal attack and their contamination with fungi often results in dramatic economic losses. Seed sterilization by plasma has been frequently examined. Basaran group has demonstrated that over 99% of grains and legume seeds exhibit at least 3 log reduction in fungal load (*Aspergillus* and *Penicillium*) after a 15 min treatment with low pressure cold plasma using air and sulfur hexafluoride (SF_6) as feeding gases, and there is no significant inhibition on seed germination (Selcuk *et al.*, 2008). Low pressure cold plasma using air and sulfur hexafluoride (SF_6) has also reduced the load of *A. parasiticus* associated with hazelnut, peanut, and pistachio nut, and SF_6 plasma is more efficient than air plasma (Basaran *et al.*, 2008). In this study, a 50% reduction in total aflatoxins (AFB1, AFB2, AFG1, and AFG2) produced by *A. parasiticus* is also observed by treatment with air plasma for 20 min, while SF_6 plasma reduced only 20% of total aflatoxin (Basaran *et al.*, 2008). Ar/O_2 Plasma jet can effectively prevent rotting of *C. fulvum* infected tomato seeds (Lu *et al.*, 2014). Treatment with dielectric barrier discharge (DBD) plasma for 10 min significantly reduces disease development in rice seeds infested with *Gibberella fujikuroi* and no adverse effects on seed germination and growth are observed (Jo *et al.*, 2014). Brassicaceous seeds contaminated with *Rhizoctonia solani* are efficiently disinfected through treatment with atmospheric pressure plasma or low pressure plasma (up to 99% reduction in infected seed number) although seed germination is delayed by atmospheric pressure plasma treatment (Nishioka *et al.*, 2014). From these experimental data, it can be suggested that plasma can become an efficient control tool for seed-borne fungal diseases without dramatic damage on seed vitality.

Antifungal activity of plasma is also observed in the treatment of fungal associated plant leaves, red pepper powder, and wood materials (Kim *et al.*, 2014; Lecoq *et al.*, 2013; Zhang *et al.*, 2014). Fungus infected spots on leaves of *Philodendron. erubescens* cv. Green Emerald are recovered after direct treatment with plasma jet on leaf spot and recovery efficiency is dependent on the size of black spots and the leaf age (Zhang *et al.*, 2014). Microwave N_2 plasma treatment reduces the number of active spores of *Aspergillus flavus* in red pepper powder by more than 2 log scale within 20 min (Kim *et al.*, 2014). Afterglows plasma can assist grafting glycidyldimethyldodecylammonium chloride (GDDAC), a known fungicidal chemical, on cellulose, which will provide antifungal properties to wood materials (Lecoq *et al.*, 2013).

The mechanism of inactivation of agriculture and food related fungi by plasma may not be remarkably different from that mentioned in plasma sterilization of medically important fungi. In most studies, fungal pathogens contaminating seeds and other agricultural products have been deactivated by plasma in open air condition and therefore reactive species generated from plasma may interact more directly with fungal pathogens. Production of diverse reactive species by using various feeding gases may enhance antifungal activity as well demonstrated in the study using SF_6 gas (Basaran *et al.*, 2008; Selcuk *et al.*, 2008). From studies to date, two interesting points about mechanisms of plasma sterilization on agricultural products can be come out. First, low pressure plasma may be more efficient in seed fungal sterilization than atmospheric pressure plasma as demonstrated in Basaran group's work (Basaran *et al.*, 2008; Selcuk *et al.*, 2008). Second, plasma may be able to efficiently inactivate fungal pathogens infected inside leaf because plasma generated species can pass through leaf stomata and damage oil vacuoles and cell membrane of fungi (Zhang *et al.*, 2014).

Conclusion

Although the limited number of studies are available compared to bacterial sterilization, studies about fungal sterilization have demonstrated that plasma can be a potential tool for efficient inactivation and disinfection of both yeast like and filamentous fungi associated with tissues, materials, air, crops, seeds, and foods. Since fungi have eukaryotic cell structure, interaction of plasma with fungal cells may be quite different from that with bacteria which are prokaryotic cells. Thus, different plasma parameters may be applicable to achieve the maximum sterilization efficiency for fungi. Various levels of antifungal activity of plasma have been observed depending on electric power, feeding gases, treatment time, fungal species, and associated environments in many studies. Control of these parameters may be closely related to modulating the level of reactive species and ions generated by plasma, and these plasma generated species may play very critical roles in fungal sterilization.

Future trends

A broad spectrum of biological effects produced by plasma have frequently been demonstrated. Plasma is likely to become an alternative sterilization tool for fungal pathogens with advantages over traditional disease control methods. In order to be an alternative antifungal tool, tremendous experimental studies *in vivo* are still needed and optimization of plasma sources for maximum efficiency should be achieved. Since fungal spores are found in various environments, finding optimal conditions for plasma application in each case is very critical for efficient control of fungal diseases and contaminants. In addition, mechanisms for plasma fungal inactivation should be actively investigated to obtain useful information in improving plasma sterilization technology.

Acknowledgements

This work was supported by the Rural Development Administration, Republic of Korea (PJ009891) and also by the National Research Foundation of Korea (NRF) (NRF-2013R1A1A3011245) and R&D Program of 'Plasma Advanced Technology for Agriculture and Food (Plasma Farming)' through the National Fusion Research Institute of Korea (NFRI) funded by the Government funds.

References

Akishev, Y., Grushin, M., Karalnik, V., Trushkin, N., Kholodenko, V., Chugunov, V., Kobzev, E., Zhirkova, N., Irkhina, I., and Kireev, G. (2008). Atmospheric-pressure, nonthermal plasma sterilization of microorganisms in liquids and on surfaces. Pure Appl. Chem. *80*, 1953–1969.

Avramidis, G., Stuwe, B., Wascher, R., Bellmann, M., Wieneke, S., von Tiedemann, A., and Viol, W. (2010). Fungicidal effects of an atmospheric pressure gas discharge and degradation mechanisms. Surf. Coat. Technol. *205*, S405-S408.

Basaran, P., Basaran-Akgul, N., and Oksuz, L. (2008). Elimination of *Aspergillus parasiticus* from nut surface with low pressure cold plasma (LPCP) treatment. Food Microbiol. *25*, 626–632.

Bennett, J.W. (2010). An overview of the genus *Aspergillus* (Caiser Academic Press, Portland).

Bennett, J.W., and Klich, M. (2003). Mycotoxins. Clin. Microbiol. Rev. *16*, 497–516.

Brown, G.D., Denning, D.W., Gow, N.A., Levitz, S.M., Netea, M.G., and White, T.C. (2012). Hidden killers: human fungal infections. Sci. Transl. Med. *4*, 165rv113.

Brun, P., Brun, P., Vono, M., Venier, P., Tarricone, E., Deligianni, V., Martines, E., Zuin, M., Spagnolo, S., Cavazzana, R., *et al.* (2012). Disinfection of ocular cells and tissues by atmospheric-pressure cold plasma. PloS one *7*, e33245.

Chew, G.L., Wilson, J., Rabito, F.A., Grimsley, F., Iqbal, S., Reponen, T., Muilenberg, M.L., Thorne, P.S., Dearborn, D.G., and Morley, R.L. (2006). Mould and endotoxin levels in the aftermath of Hurricane Katrina: a pilot project of homes in New Orleans undergoing renovation. Environ. Health Perspect. *114*, 1883–1889.

Cortez, K.J., Roilides, E., Quiroz-Telles, F., Meletiadis, J., Antachopoulos, C., Knudsen, T., Buchanan, W., Milanovich, J., Sutton, D.A., Fothergill, A., *et al.* (2008). Infections caused by *Scedosporium* spp. Clin. Microbiol. Rev. *21*, 157–197.

Czylkowski, D., Hrycak, B., Jasinski, M., Dors, M., and Mizeraczyk, J. (2013). Atmospheric pressure microwave microplasma microorganism deactivation. Surf. Coat. Technol. *234*, 114–119.

Daeschlein, G., Scholz, S., von Woedtke, T., Niggemeier, M., Kindel, E., Brandenburg, R., Weltmann, K.D., and Junger, M. (2011). In vitro killing of clinical fungal strains by low-temperature atmospheric-pressure plasma jet. IEEE Trans. Plasma Sci. *39*, 815–821.

Donnelly, J.P. (2013). Care pathways for managing invasive mould diseases. Int. J. Antimicrob. Agents *42*, 5–9.

Fisher, M.C., Henk, D.A., Briggs, C.J., Brownstein, J.S., Madoff, L.C., McCraw, S.L., and Gurr, S.J. (2012). Emerging fungal threats to animal, plant and ecosystem health. Nature *484*, 186–194.

Fridman, G., Friedman, G., Gutsol, A., Shekhter, A.B., Vasilets, V.N., and Fridman, A. (2008). Applied plasma medicine. Plasma Process. Polym. *5*, 503–533.

Grunwald, N.J., Goss, E.M., and Press, C.M. (2008). *Phytophthora ramorum*: a pathogen with a remarkably wide host range causing sudden oak death on oaks and ramorum blight on woody ornamentals. Mol. Plant Pathol. *9*, 729–740.

Hedayati, M.T., Pasqualotto, A.C., Warn, P.A., Bowyer, P., and Denning, D.W. (2007). *Aspergillus flavus*: human pathogen, allergen and mycotoxin producer. Microbiol. *153*, 1677–1692.

Hirooka, T., and Ishii, H. (2013). Chemical control of plant diseases. J. Gen. Plant Pathol. *79*, 390–401.

Iseki, S., Ohta, T., Aomatsu, A., Ito, M., Kano, H., Higashijima, Y., and Hori, M. (2010). Rapid inactivation of *Penicillium digitatum* spores using high-density nonequilibrium atmospheric pressure plasma. Appl. Phys. Lett. *96*, 153704.

Ishii, H. (2006). Impact of fungicide resistance in plant pathogens on crop disease control and agricultural environment. Jpn. Agric. Res. Q.: JARQ *40*, 205–211.

Ito, S., Ihara, T., Tamura, H., Tanaka, S., Ikeda, T., Kajihara, H., Dissanayake, C., Abdel-Motaal, F.F., and El-Sayed, M.A. (2007). Alpha-tomatine, the major saponin in tomato, induces programmed cell death mediated by reactive oxygen species in the fungal pathogen *Fusarium oxysporum*. FEBS Lett. *581*, 3217–3222.

Jo, Y.K., Cho, J., Tsai, T.C., Staack, D., Kang, M.H., Roh, J.H., Shin, D.B., Cromwell, W., and Gross, D. (2014). A Non-thermal plasma seed treatment method for management of a seedborne fungal pathogen on rice seed. Crop Sci. *54*, 796–803.

Kim, J.E., Lee, D.U., and Min, S.C. (2014). Microbial decontamination of red pepper powder by cold plasma. Food Microbiol. *38*, 128–136.

Korachi, M., Turan, Z., Senturk, K., Sahin, F., and Aslan, N. (2009). An investigation into the biocidal effect of high voltage AC/DC atmospheric corona discharges on bacteria, yeasts, fungi and algae. J. Electrostat. *67*, 678–685.

Lecoq, E., Desbrieres, J., Clement, F., Leclaire, C., and Held, B. (2013). Wood protective fungicidal treatment: quaternary ammonium molecules grafting on cellulose assisted by afterglows at atmospheric pressure. Plasma Process. Polym. *10*, 150–160.

Lu, Q., Liu, D., Song, Y., Zhou, R., and Niu, J. (2014). Inactivation of the tomato pathogen *Cladosporium fulvum* by an atmospheric-pressure cold plasma jet. Plasma Process. Polym. *11*, 1028–1036.

Mizeraczyk, J., Dors, M., Jasinski, M., Hrycak, B., and Czylkowski, D. (2013). Atmospheric pressure low-power microwave microplasma source for deactivation of microorganisms. Eur. Phys. J. Appl. Phys. *61*, 24309.

Na, Y.H., Park, G., Choi, E.H., and Uhm, H.S. (2013). Effects of the physical parameters of a microwave plasma jet on the inactivation of fungal spores. Thin Solid Films *547*, 125–131.

Nakamura, Y., Utsumi, Y., Suzuki, N., Nakajima, Y., Murata, O., Sasaki, N., Nitanai, H., Nagashima, H., Miyamoto, S., Yaegashi, J., *et al.* (2011). Multiple *Scedosporium apiospermum* abscesses in a woman survivor of a tsunami in northeastern Japan: a case report. J. Med. Case Rep. *5*, 526.

Nishioka, T., Takai, Y., Kawaradani, M., Okada, K., Tanimoto, H., Misawa, T., and Kusakari, S. (2014). Seed disinfection effect of atmospheric pressure plasma and low pressure plasma on *Rhizoctonia solani*. Biocontrol Sci. *19*, 99–102.

Nojima, H., Park, R.-E., Kwon, J.-H., Suh, I., Jeon, J., Ha, E., On, H.-K., Kim, H.-R., Choi, K., Lee, K.-H., *et al.* (2007). Novel atmospheric pressure plasma device releasing atomic hydrogen: reduction of microbial-contaminants and OH radicals in the air. J. Phys. D: Appl. Phys. *40*, 501.

Odds, F.C. (1987). *Candida* infections: an overview. Crit. Rev. Microbiol. *15*, 1–5.

Ohkawa, H., Akitsu, T., Tsuji, M., Kimura, H., and Fukushima, K. (2005). Initiation and microbial-disinfection characteristics of wide-gap atmospheric-pressure glow discharge using soft X-ray ionization. Plasma Process. Polym. *2*, 120–126.

Ohkawa, H., Akitsu, T., Tsuji, M., Kimura, H., Kogoma, M., and Fukushima, K. (2006). Pulse-modulated, high-frequency plasma sterilization at atmospheric-pressure. Surf. Coat. Technol. *200*, 5829–5835.

Panngom, K., Lee, S.H., Park, D.H., Sim, G.B., Kim, Y.H., Uhm, H.S., Park, G., and Choi, E.H. (2014). Non-thermal plasma treatment diminishes fungal viability and up-regulates resistance genes in a plant host. PloS one *9*, e99300.

Park, B.J., Lee, D.H., Park, J.C., Lee, I.S., Lee, K.Y., Hyun, S.O., Chun, M.S., and Chung, K.H. (2003). Sterilization using a microwave-induced argon plasma system at atmospheric pressure. Phys. Plasmas *10*, 4539–4544.

Park, B.J., Takatori, K., Sugita-Konishi, Y., Kim, I.H., Lee, M.H., Han, D.W., Chung, K.H., Hyun, S.O., and Park, J.C. (2007). Degradation of mycotoxins using microwave-induced argon plasma at atmospheric pressure. Surf. Coat. Technol. *201*, 5733–5737.

Park, D.J., Lee, M.H., Woo, Y.I., Han, D.-W., Choi, J.B., Kim, J.K., Hyun, S.O., Chung, K.-H., and Park, J.-C. (2008). Sterilization of microorganisms in silk fabrics by microwave-induced argon plasma treatment at atmospheric pressure. Surf. Coat. Technol. *202*, 5773–5778.

Park, J.C., Park, B.J., Han, D.W., Lee, D.H., Lee, I.S., Hyun, S.O., Chun, M.S., Chung, K.H., Aihara, M., and Takatori, K. (2004). Fungal sterilization using microwave-induced argon plasma at atmospheric pressure. J. Microbiol. Biotechnol. *14*, 188–192.

Pfaller, M.A., and Diekema, D.J. (2005). Unusual fungal and pseudofungal infections of humans. J. Clin. Microbiol. *43*, 1495–1504.

Rao, C.Y., Riggs, M.A., Chew, G.L., Muilenberg, M.L., Thorne, P.S., Van Sickle, D., Dunn, K.H., and Brown, C. (2007). Characterization of airborne moulds, endotoxins, and glucans in homes in New Orleans after hurricanes Katrina and Rita. Appl. Environ. Microbiol. *73*, 1630–1634.

Rypien, K.L., Andras, J.P., and Harvell, C.D. (2008). Globally panmictic population structure in the opportunistic fungal pathogen *Aspergillus sydowii*. Mol. Ecol. *17*, 4068–4078.

Sawada, H., Sugihara, M., Takagaki, M., and Nagayama, K. (2004). Monitoring and characterization of Magnaporthe grisea isolates with decreased sensitivity to scytalone dehydratase inhibitors. Pest Manag. Sci. *60*, 777–785.

Sedghizadeh, P.P., Meng-Tse, C., Schaudinn, C., Gorur, A., and Chunqi, J. (2012). Inactivation kinetics study of an atmospheric-pressure cold-plasma jet against pathogenic microorganisms. IEEE Trans. Plasma Sci. *40*, 2879–2882.

Selcuk, M., Oksuz, L., and Basaran, P. (2008). Decontamination of grains and legumes infected with *Aspergillus* spp. and *Penicillum* spp. by cold plasma treatment. Bioresour. Technol. *99*, 5104–5109.

Shor, E., and Perlin, D.S. (2015). Coping with stress and the emergence of multidrug resistance in fungi. PLoS Pathog. *11*, e1004668.

Singh, R.P., Hodson, D.P., Huerta-Espino, J., Jin, Y., Bhavani, S., Njau, P., Herrera-Foessel, S., Singh, P.K., Singh, S., and Govindan, V. (2011). The emergence of Ug99 races of the stem rust fungus is a threat to world wheat production. Annu. Rev. Phytopathol. *49*, 465–481.

Song, Y., Liu, D., Ji, L., Wang, W., Zhao, P., Quan, C., Niu, J., and Zhang, X. (2012). The inactivation of resistant *Candida albicans* in a sealed package by cold atmospheric pressure plasmas. Plasma Process. Polym. *9*, 17–21.

Song, Y., Wang, W., Lu, Q., Yang, D., Niu, J., and Liu, D. (2013). A homogeneous surface inactivation device driven by a pulse high-voltage source. Plasma Process. Polym. *10*, 698–705.

Thornton, C.R., and Wills, O.E. (2015). Immunodetection of fungal and oomycete pathogens: established and emerging threats to human health, animal welfare and global food security. Crit. Rev. Microbiol. *41*, 27–51.

Vrajova, J., Chalupova, L., Novotny, O., Cech, J., Krcma, F., and Stahel, P. (2008). Removal of paper microbial contamination by atmospheric pressure DBD discharge. Eur. Phys. J. D *54*, 233–237.

Wilkinson, L. (1988). Introduction to the history of medical and veterinary mycology. Med. Hist. *32*, 102–103.

Wilson, R.A., and Talbot, N.J. (2009). Under pressure: investigating the biology of plant infection by *Magnaporthe oryzae*. Nat. Rev. Micro. *7*, 185–195.

Xiong, Z., Lu, X.P., Feng, A., Pan, Y., and Ostrikov, K. (2010). Highly effective fungal inactivation in He+ O_2 atmospheric-pressure nonequilibrium plasmas. Phys. Plasmas *17*, 123502.

Zhang, X., Liu, D., Zhou, R., Song, Y., Sun, Y., Zhang, Q., Niu, J., Fan, H., and Yang, S. (2014). Atmospheric cold plasma jet for plant disease treatment. Appl. Phys. Lett. *104*, 043702.

Current Progress in Advanced Research into the Inactivation of Viruses by Gas Plasma: Influenza Virus Inactivation by Nitrogen Gas Plasma

10

Akikazu Sakudo

Abstract

Recently, the potential of gas plasma technology for disinfection and sterilization has been exploited. The effective inactivation of bacteria, fungi as well as toxins by gas plasma exemplifies the broad application of this technology. However, information concerning the effect of gas plasma on viruses and their biomolecules, as well its mechanism of action, remain somewhat limited. In this chapter, we focus on the inactivation of viruses by gas plasma, together with our recent investigation on inactivation of influenza viruses by nitrogen gas plasma, which was produced by applying a short high-voltage pulse using a static induction (SI) thyristor power supply.

Introduction

Iatrogenic diseases caused by infection through contact with medical devices are thought to be a major contributor to hospital-acquired infections (Fraise *et al.*, 2013). Consequently, the appropriate sterilization of medical devices is crucial for decreasing the incidence of iatrogenic diseases. However, medical devices and instruments are often not resistant to heat treatment such as autoclaving and dry-heating (McCombs and Darby, 2010). Moreover, alternative sterilization techniques involving the generation of γ-rays or electron beams require expensive facilities and are not appropriate for routine use (Silindir and Ozer, 2012). Ethylene oxide gas (EOG) can be used to sterilize heat-sensitive medical instruments, but this approach has limited application because the gas is both toxic and carcinogenic (Greim, 2003). Recently, sterilization using hydrogen peroxide gas or its gas plasma was proposed, although this technique is ineffective against endotoxins and lipopolysaccharides (LPSs) (Tamazawa and Hashibuchi, 2004; Shintani *et al.*, 2007). Indeed, residual amounts of endotoxin derived from bacteria may cause symptoms including fever (Blattais, 2006). By contrast, there is growing evidence to suggest that gas plasma technology is a valuable method for both disinfection and sterilization.

Recently, the number of studies utilizing plasma technology has been increasing (Machala *et al.*, 2012). Diverse applications of gas plasma have been reported in the fields of medical and clinical science (Terrier *et al.*, 2009; Isbary *et al.*, 2013; Hoffmann *et al.*, 2013;

Sakudo *et al.*, 2014), dentistry (Hoffmann *et al.*, 2013), materials processing (Abuzairi *et al.*, 2015; Law *et al.*, 2012), material analysis (Sato *et al.*, 2014), surface modification (Sakudo *et al.*, 2015a,b; Saraswati *et al.*, 2013), light sources (Sato *et al.*, 2014; Jinno *et al.*, 2006), and microplasma chips (Topala and Nagatsu, 2015), as well as the food industry (Banu *et al.*, 2012; Pankaj *et al.*, 2014; Maeda *et al.*, 2015; Afshari and Hosseini, 2014). Among them, over the past 20 years, many researchers have shown an interest in a disinfection technique based on gas plasma technology (Laroussi, 1996; Laroussi *et al.*, 2012; Fridman and Friedman, 2013). Recently, *Escherichia coli*, *Bacillus* species, *Salmonella*, *Staphylococcus aureus*, *Enterococcus faecalis*, and *Penicillium digitatum*, in addition to bacterial spores such as *Geobacillus stearothermophilus*, have been shown to be inactivated by gas plasma treatment (Niemira, 2012; Klämpfl *et al.*, 2012; Sung *et al.*, 2013; Tian *et al.*, 2010; Iseki *et al.*, 2010), indicating that gas plasma technology is highly effective for disinfecting bacteria. However, there have only been a small number of studies on the inactivation of viruses such as influenza virus (Sakudo *et al.*, 2013a), adenovirus (Zimmermann *et al.*, 2011), and feline calicivirus (Aboubakr *et al.*, 2015).

As mentioned above, the effective inactivation of bacteria and fungi as well as toxins by gas plasma has been demonstrated, suggesting the broad application of this technology. However, information concerning the effect of gas plasma on viruses and their biomolecules, as well its mechanism of action, remain somewhat limited. In this chapter, we focus on the inactivation of viruses by gas plasma, together with our recent investigation on the inactivation of influenza viruses by nitrogen gas plasma.

Viruses and influenza virus

Viruses are divided into two types; enveloped and non-enveloped viruses (Fig. 10.1). Enveloped viruses have a lipid membrane on their outer surface, whereas non-enveloped viruses lack such a membrane. Currently, most studies on virus inactivation by gas plasma have been performed using bacteriophages, which specifically infect bacterial cells and do not cause iatrogenic infections.

Influenza viruses are enveloped viruses. Human influenza viruses are a major causative

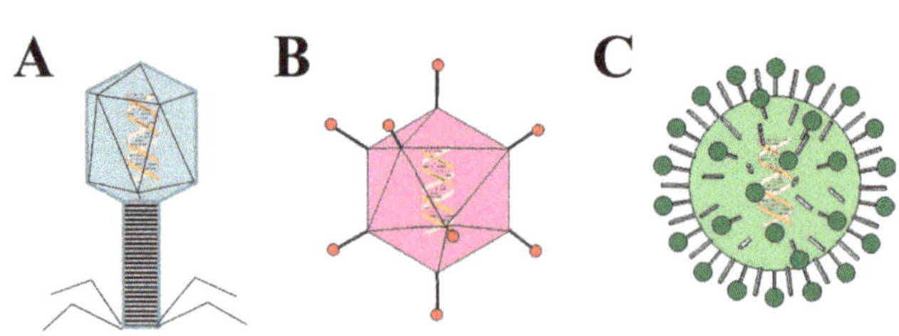

Figure 10.1 Animal viruses and bacteriophages. Schematic structures of λ bacteriophage (A), non-enveloped virus (B) and enveloped virus (C) are shown. Bacteriophages infect bacteria. Animal viruses infect animal cells and can be divided into non-enveloped and enveloped viruses. Non-enveloped animal viruses includes adenovirus, norovirus and rotavirus, while animal enveloped viruses includes influenza virus, human immunodeficiency virus (HIV) and respiratory syncytial virus (RSV). To date, investigations into the plasma sterilization of viruses have mainly focused on bacteriophages, while there are a limited number of studies examining animal viruses.

agent of severe upper respiratory tract infections in infants, young children and the elderly. It is estimated that infection with influenza virus results in 3 to 5 million cases of severe illness and 250,000 to 500,000 deaths per year worldwide (WHO, 2003). As such, inactivation of influenza virus is a major objective for improving human health. Recently, we have investigated whether gas plasma can be used for the efficient sterilization or disinfection of influenza viruses. Specifically, we investigated the virucidal effect of nitrogen gas plasma treatment as well as any associated biochemical changes to the components of the influenza virus. Next, I will introduce these recent studies using the nitrogen gas plasma instrument.

Inactivation of influenza virus by nitrogen gas plasma

Recently, the effect of nitrogen gas plasma on viruses has been studied by several research groups including ourselves. Here, we review our recent studies on nitrogen gas plasma treatment of influenza virus, as a representative enveloped virus. In addition, the mechanisms by which the influenza virus is inactivated by nitrogen gas plasma is discussed.

We have used nitrogen gas plasma produced by applying a short high-voltage pulse using a static induction (SI) thyristor power supply (Fig. 10.2). The instrument is referred to as BLP-TES. Influenza virus-infected allantoic fluid was spotted onto a cover glass and subjected to nitrogen gas plasma treatment. The appearance of the spots was unchanged after nitrogen gas plasma treatment for up to 15 min (Fig. 10.3).

To further investigate the change induced by nitrogen gas plasma, the treated influenza virus was analysed using a number of different assays. A bioassay using chicken embryonated eggs demonstrated that nitrogen gas plasma treated influenza virus could not proliferate,

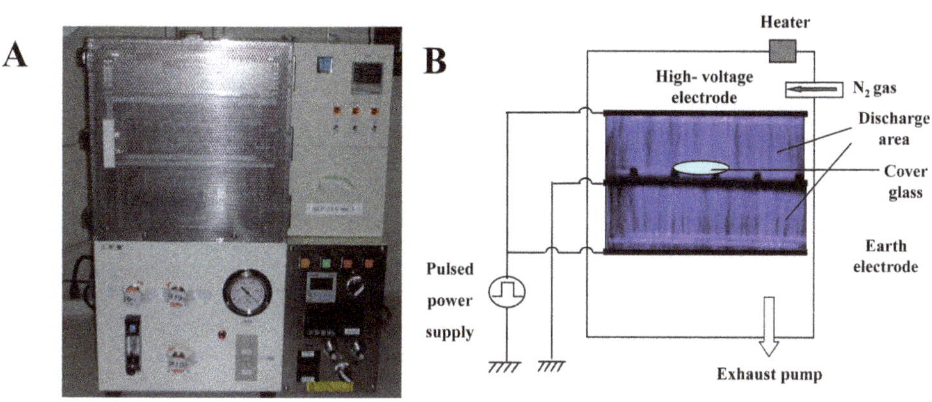

Figure 10.2 Nitrogen gas plasma instrument BLT-TE. (A) Photograph of the nitrogen gas plasma instrument (BLP-TES). BLP-TES produces nitrogen gas plasma by means of a fast high-voltage pulse applied using a static induction (SI) thyristor power supply. A cathode electrode (earth electrode) is placed between the anode electrodes (high-voltage electrode). (B) Schematic of the nitrogen gas plasma instrument. The distance between the high-voltage electrode and the earth electrode is 50 mm. The procedure for generating the nitrogen gas plasma was as follows. First, the chamber box containing the sample was decompressed and degassed, and then nitrogen gas (99.9995%, Okano, Okinawa, Japan) was introduced. The flow rate of nitrogen gas was 10 l/min. The pressure in the box was maintained at about 0.5 atmospheres during the discharge at 1.5 kpps (kilopulse per second). Modified from Figure 1 in Sakudo *et al.* (2014).

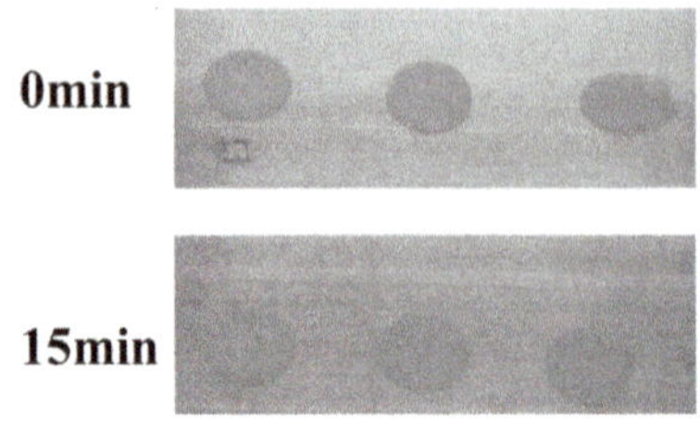

Figure 10.3 No change in the appearance of allantoic fluid infected with influenza virus after nitrogen gas plasma treatment. Influenza A virus (A/PR/8/34)-infected allantoic fluid was spotted onto a cover glass and air-dried. The cover glass was then treated with nitrogen gas plasma using the BLP-TES instrument at 1.5 kpps for 15 min. The appearance of the spots was unaffected after nitrogen gas plasma treatment for up to 15 min by comparison with the untreated spots (0 min).

suggesting inactivation of the virus during treatment (Fig. 10.4). Scanning electron microscopy (SEM) analysis showed aggregation and fusion of viruses in the nitrogen gas plasma treated samples (Fig. 10.5). Moreover, nitrogen gas plasma treatment of influenza A and B viruses was shown to induce the degradation of viral proteins, including NP, HA and NA, as shown by immunochromatography for influenza virus nucleoprotein (NP), Coomassie brilliant blue staining for proteins, enzyme-linked immunosorbent assay (ELISA) for influenza virus NP, haemagglutination assay for haemagglutinin (HA), and Western blotting for influenza virus NP and neuraminidase (NA). Polymerase chain reaction (PCR) analysis suggested that the viral RNA genome was damaged and may be oxidized by nitrogen gas plasma treatment (Fig. 10.6). Fourier transform infrared (FT-IR) spectroscopy analysis indicated that nitrogen gas plasma treatment induces changes to the lipids, suggesting that

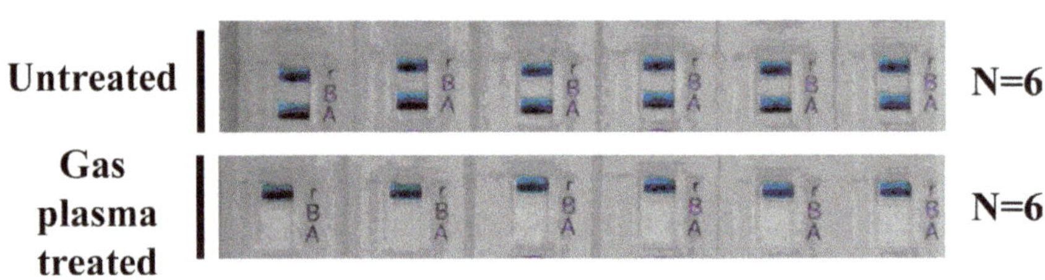

Figure 10.4 Inactivation of influenza A virus by nitrogen gas plasma occurs within 5 min. A 20 µl aliquot of allantoic fluid [3×10^{14} $TCID_{50}$ (tissue culture infectious dose 50) per ml] infected with influenza A virus (A/PR/8/34) was spotted and air-dried onto a cover glass and then treated with nitrogen gas plasma (1.5 kpps) for 5 min using BLP-TES. Each sample was recovered with 20 µl of pure water (Otsuka Pharmaceuticals, Co., Tokyo, Japan) and injected into 11-day-old chicken embryonated eggs. Specifically, six eggs were injected with recovered solution from the nitrogen gas plasma-treated allantoic fluid, while a further six eggs were injected with recovered solution from untreated allantoic fluid. After incubation of the eggs for 48 h at 37°C, the presence of influenza virus was analysed by immunochromatography for influenza virus nucleoprotein (NP) (ESPRINE Influenza A&B-N; Fujirebio Inc., Tokyo, Japan). The result of these experiments indicated that nitrogen gas plasma treatment (1.5 kpps) effectively inactivated influenza virus within 5 min. A and B indicate the lines for NP of influenza A virus and influenza B virus, respectively. The reference line (r) is also indicated. Modified from Figure 2 in Sakudo *et al.* (2013).

Figure 10.5 Morphological changes to influenza viruses treated with nitrogen gas plasma were observed by scanning electron microscopy (SEM). Influenza A virus (A/PR/8/34) in allantoic fluid was air-dried on a cover glass and treated for 5 min with an nitrogen gas plasma (1.5 kpps) generated using a BLP-TES, and then fixed with 2% glutaraldehyde/0.1 M phosphate buffer (pH 7.4). The cover glasses were then treated with 2% osmium tetroxide, dehydrated in ethanol, dried to the critical point and finally coated with osmium plasma ions. Virus morphology was analysed by SEM at 5 kV using a magnification of ×100,000 with a JSM-6320F instrument (JEOL Ltd., Tokyo, Japan). Untreated influenza A viruses displayed a rounded shape with fibrous connections, whereas nitrogen gas plasma-treated influenza A viruses had a shrunken appearance and lacked any associated fibres. Reproduced from Figure 3 in Sakudo *et al*. (2014).

the viral envelope and its associated lipids become oxidized. In addition, viral proteins were nitrated after nitrogen gas plasma treatment (Fig. 10.7). Based on molecular weight information, the main viral protein to be nitrated appears to be NA. In terms of the mechanism by which nitrogen gas plasma inactivates influenza viruses, there are thought to be at least three main factors. These inactivation factors include heat, longwave ultraviolet A (UV-A), and oxidative stress (i.e. exposure to hydrogen peroxide-like molecules). Further calculation of the contribution ratio of each of these factors on the inactivation of influenza virus suggests that oxidative stress is the main mechanism of inactivation. However, it remains unclear how other inactivation factors contribute to the observed virucidal effect. Additional oxidative stress factors, such as the generation of reactive chemical products other than hydrogen peroxide, may be involved in viral inactivation. Further studies are required to identify these potential inactivation factors.

Conclusion and future perspectives

In conclusion, the inactivation effect of gas plasma on viruses has been confirmed. Our analysis shows that reactive chemical products generated by gas plasma, such as hydrogen peroxide, induce oxidative stress and this is chiefly responsible for the viral inactivation. Currently, gas plasma research has largely focused on potential medical applications such as sterilization, disinfection and antisepsis, including the inactivation of influenza viruses. However, recently, the potential application of gas plasma treatment in the agricultural sector is also being investigated. Although information concerning the effect of gas plasma

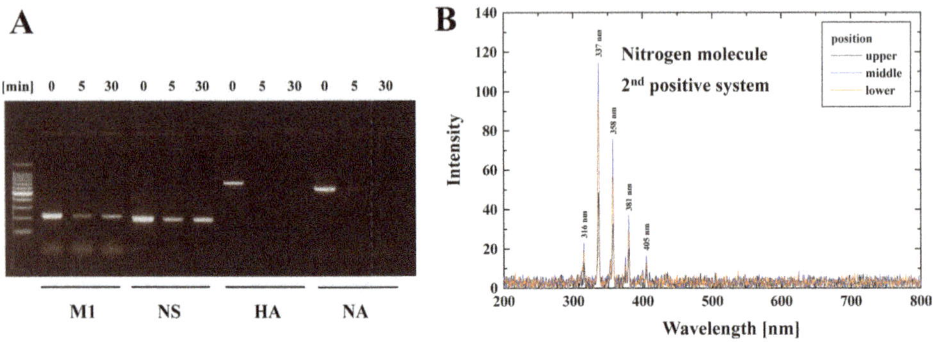

Figure 10.6 Nitrogen gas plasma generates long wave ultraviolet A (UV-A) and induces damage to viral genomic RNA of influenza virus. (A) An aliquot (20 µl) of influenza A virus (A/PR/8/34)-infected allantoic fluid was dropped onto a cover glass and air-dried. The dried spot was then treated with nitrogen gas plasma (1.5 kpps) for the indicated times using BLP-TES. The viral genomic RNA of the influenza virus was subsequently extracted and subjected to reverse transcription in order to generate complementary DNA (cDNA). Damage to the genomic RNA of influenza virus induced by the nitrogen gas plasma treatment was analysed using the polymerase chain reaction (PCR). Namely, PCR was performed using cDNAs as template and specific sets of primers for matrix protein (M1), non-structural protein (NS), haemagglutinin (HA) and neuraminidase (NA) of influenza virus. PCR was carried out using the following primers; MA1-F: 5′-CAG AGA CTT GAA GAT GTC TTT GCT G-3′; MA1-R: GCT CTG TCC ATG TTA TTT GGA TC-3′; HA-F: 5′-AGC AAA AGC AGG GGA AAA TAA –3′; HA-R: 5′-GCT ATT TCT GGG GTG AAT CT-3′; NA-F: 5′-TTG CTT GGT CGG CAA GTG C-3′; NA-R: 5′-CCA GTC CAC CCA TTT GGA TCC-3′;;NS-F: 5′-AAG GGC TTT CAC CGA AGA GG-3′; NS-R: 5′-CCC ATT CTC ATT ACT GCT TC-3′. The temperature cycling conditions used for the PCR were: 95°C for 5 min followed by 25 cycles of 95°C for 1 min, 55°C for 1 min and 72°C for 1 min with one final cycle of 72°C for 10 min. The intensity of the resulting amplified bands was semi-quantitatively analysed by agarose gel electrophoresis. Bands in test samples were visually compared to those in untreated controls. The amplified PCR products generated from each pair of primers were verified by DNA sequencing. DNA size marker (100 bp DNA ladder Dye plus, Takara Bio Inc.) was run on the left-hand lane of the gel. (B) Emission of radiation between 200 and 800 nm was analysed using a UV-Vis-Near-infrared multichannel spectrophotometer (S-2431; Soma Optics Ltd., Tokyo, Japan). The intensity of emission at each wavelength collected from the upper, middle, and lower position of the discharge region using an optical fibre probe is shown. Peaks related to the second positive system of nitrogen (316, 337, 381, 405 nm) were detected. Modified from Figure 8 in Sakudo *et al.* (2013).

treatment on plant related bacteria and fungi has accumulated, there is a lack of data for plant viruses. As such, there are no reports describing the effectiveness of gas plasma treatment on plant viruses. Therefore, research needs to be focused on this area if the technology is to be applied to the disinfection of agriculturally related viruses.

Acknowledgements

This work was supported by Grant-in-Aid for Science and Technology research promotion programme for agriculture, forestry, fisheries and food industry. Some of the data presented in this chapter has been previously reported (Sakudo *et al.*, 2013; Sakudo *et al.*, 2014). Hindawi Publishing Corporation and Frontiers in Bioscience, which are the publishers of the previous papers, have granted permission to all authors to use the figures that appear in their own papers.

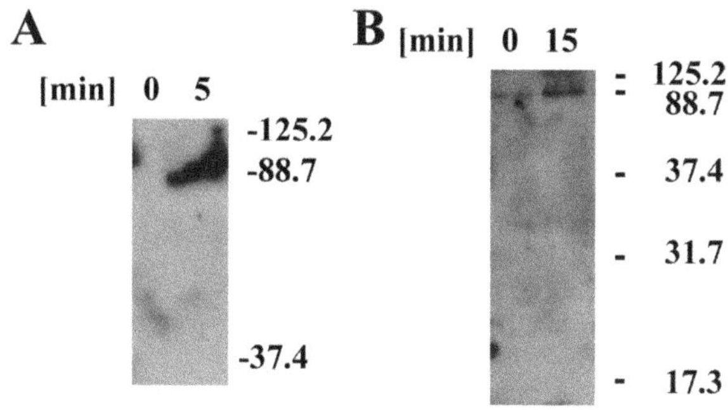

Figure 10.7 Nitrated proteins in influenza virus-infected allantoic fluid were induced by nitrogen gas plasma treatment. Proteins in influenza virus (A/PR/8/34)-infected allantoic fluid (3×10^{14} $TCID_{50}$/ml) treated with nitrogen gas plasma (1.5 kpps) using BLP-TES for 0 and 5 min (A) or 0 and 15 min (B) were subjected to sodium dodecyl sulfate (SDS)-polyacrylamide gel electrophoresis (PAGE) and transblotted onto a polyvinylidene fluoride (PVDF) membrane. Nitrated proteins on the membrane were detected with anti-nitrotyrosine antibody (MAB5404; Millipore, Billerica, MA). The amount of nitrated proteins increased after nitrogen gas plasma treatment. The molecular mass marker (kDa) is shown on the right hand side. The intensely stained band with a molecular mass of ~88.7 kDa may correspond to nitrated NA. Reproduced from Figure 10 in Sakudo *et al.* (2014).

References

Aboubakr, H.A., Williams, P., Gangal, U., Youssef, M.M., El-Sohaimy, S.A., Bruggeman, P.J., and Goyal, S.M. (2015). Virucidal effect of cold atmospheric gaseous plasma on feline calicivirus, a surrogate for human norovirus. Appl. Environ. Microbiol. *81*, 3612–3622.

Abuzairi, T., Okada, M., Mochizuki, Y., Poespawati, N.R., Wigajatri, R., Nagatsu, M., 2015. Maskless functionalization of a carbon nanotube dot array biosensor using an ultrafine atmospheric pressure plasma jet. Carbon *89*. 208–216.

Afshari, R., and Hosseini, H. (2014). Non-thermal plasma as a new food preservation method, Its present and future prospect. J. Paramed. Sci. *5*, 2008–4978.

Banu, S.M., Sasikala, P., Dhanapal, A., Kavitha, V., Yazhini, G., and Rajamani, L. (2012). Cold plasma as a novel food processing technology. Int. J. Emerg. Trend Eng. Dev. *4*, 803–818.

Blatteis, C.M. (2006). Endotoxic fever: new concepts of its regulation suggest new approaches to its management. Pharmacol. Ther. *111*, 194–223.

Fraise, A., Maillard, J.Y., and Sattar, S. (2013). Russell, Hugo and Ayiffe's Principles and Practice of Disinfection Preservation and Sterilization (New Jersey, USA: Wiley-Blackwell).

Fridman, A., and Friedman, G., 2013. Plasma Medicine (New Jersey, USA: Wiley-Blackwell).

Greim, H. (2003). Mechanistic and toxicokinetic data reducing uncertainty in risk assessment. Toxicol. Lett. *138*, 1–8.

Hoffmann, C., Berganza, C., and Zhang, J. (2013). Cold atmospheric plasma: methods of production and application in dentistry and oncology. Med. Gas Res. *3*, 21.

Isbary, G., Shimizu, T., Zimmermann, J.L., Thomas, H.M., Morfill, G.E., and Stolz, W. (2013). Cold atmospheric plasma for local infection control and subsequent pain reduction in a patient with chronic post-operative ear infection. New Microbes New Infect. *1*, 41–43.

Iseki, S., Ohta, T., Aomatsu, A., Ito, M., Kano, H., Higashijima, Y., and Hori, M. (2010). Rapid inactivation of *Penicillium digitatum* spores using high-density nonequilibrium atmospheric pressure plasma. Appl. Phys. Lett. *96*, 153704.

Jinno, M., Taniuchi, H., Watanabe, M., and Motomura, H. (2006). Optimal Pulse voltage waveform for a xenon barrier discharge lamp using both an inner electrode and an external electrode. J. Light Vis. Environ. *30*, 58–60.

Klampfl, T.G., Isbary, G., Shimizu, T., Li, Y.F., Zimmermann, J.L., Stolz, W., Schlegel, J., Morfill, G.E., and Schmidt, H.U. (2012). Cold atmospheric air plasma sterilization against spores and other microorganisms of clinical interest. Appl. Environ. Microbiol. *78*, 5077–5082.

Laroussi, M., Kong, M.G., Morfill, G., and Stolz, W. (2012). Plasma Medicine: Application of low-Temperature Gas Plasmas in Medicine and Biology (Cambridge, UK, Cambridge University Press).

Laroussi, M. (1996). Sterilization of Contaminated Matter with an Atmospheric Pressure Plasma. IEEE Trans. Plasma Sci. *24*, 1188–1191.

Law, V.J., O'Neill, F.T., Twomey, B., Milosavljevi, V., Kong, M.G., Anghel, S.D., and Dowling, D.P. (2012). Electrical power dissipation within a helium APPJ flowing afterglow and its impact on spatial-temporal properties. IEEE Trans. Plasma Sci. *40*, 2994–3002.

McCombs, G.B., and Darby, M.L. (2010). New discoveries and directions for medical, dental and dental hygiene research: low temperature atmospheric pressure plasma. Int. J. Dent. Hyg. *8*, 10–15.

Machala, Z., Hensel, K., and Akishev, Y. (2012). Plasma for Bio-Decontamination, Medicine and Food Security (NATO Science for Peace and Security Series A: Chemistry and Biology) (New York, Springer).

Maeda, K., Toyokawa, Y., Shimizu, N., Imanishi, Y., and Sakudo, A. (2015). Inactivation of *Salmonella* by nitrogen gas plasma generated by a static induction thyristor as a pulsed power supply. Food Control *52*, 54–59.

Niemira, B.A. (2012). Cold plasma reduction of *Salmonella* and *Escherichia coli* O157:H7 on almonds using ambient pressure gases. J. Food Sci. *77*, M171–175.

Pankaj, S.K., Bueno-Ferrer, C., Misra, N.N., Milosavljevi, V., O'Donnell, C.P., Bourke, P., Keener, K.M., and Cullen, P.J. (2014). Applications of cold plasma technology in food packaging, Trend. Food Sci. Technol. *35*, 5–17.

Sakudo, A., Chou, H., and Nagatsu, M. (2015a). Antibody-integrated and functionalized graphite-encapsulated magnetic beads, produced using ammonia gas plasma technology, for capturing *Salmonella*. Bioorg. Med. Chem. Lett. *25*, 1012–1016.

Sakudo, A., Chou, H., Ikuta, K., and Nagatsu, M. (2015b). Integration of antibody by surface functionalization of graphite-encapsulated magnetic beads using ammonia gas plasma technology for capturing influenza A virus. Bioorg. Med. Chem. Lett. *25*, 1876–1879.

Sakudo, A., Misawa, T., Shimizu, N., and Imanishi, Y. (2014). N(2) gas plasma inactivates influenza virus mediated by oxidative stress. Front. Biosci. (Elite Ed.) *6*, 69–79.

Sakudo, A., Shimizu, N., Imanishi, Y., and Ikuta, K. (2013a). N_2 gas plasma inactivates influenza virus by inducing changes in viral surface morphology, protein, and genomic RNA. Biomed. Res. Int. *2013*, 694269.

Saraswati, T.E., Tsumura, S., and Nagatsu, M. (2014). High-efficiency plasma surface modification of graphite-encapsulated magnetic nanoparticles using a pulsed particle explosion technique. Jpn. J. Appl. Phys. *53*, 010205.

Sato, R., Yasumatsu, D., Kumagai, S., Takeda, K., Hori, M., and Sasaki, M. (2014). An atmospheric pressure inductively coupled microplasma source of vacuum ultraviolet light. Sens. Act. A-Phys. *215*, 144–149.

Shintani, H., Shimizu, N., Imanishi, Y., Sekiya, T., Tamazawa, K., Taniguchi, A., and Kido, N. (2007). Inactivation of microorganisms and endotoxins by low temperature nitrogen gas plasma exposure. Biocontrol Sci. *12*, 131–143.

Silindir, M., and Ozer, Y. (2012). The effect of radiation on a variety of pharmaceuticals and materials containing polymers, PDA journal of pharmaceutical science and technology. PDA *66*, 184–199.

Sung, S.J., Huh, J.B., Yun, M.J., Chang, B.M., Jeong, C.M., and Jeon, Y.C. (2013). Sterilization effect of atmospheric pressure non-thermal air plasma on dental instruments. J. Adv. Prosthodont. *5*, 2–8.

Tamazawa, K., and Hoshibuchi, K. (2004). Development of endotoxin inactivated by oxygen plasma. Jpn. J. Med. Inst. *74*, 285–291.

Terrier, O., Essere, B., Yver, M., Barthelemy, M., Bouscambert-Duchamp, M., Kurtz, P., VanMechelen, D., Morfin, F., Billaud, G., Ferraris, O., Lina, B., Rosa-Calatrava, M., and Moules, V. (2009). Cold oxygen plasma technology efficiency against different airborne respiratory viruses. J. Clin. Virol. *45*, 119–124.

Tian, Y., Sun, P., Wu, H., Bai, N., Wang, R., Zhu, W., Zhang, J., and Liu, F. (2010). Inactivation of *Staphylococcus aureus* and *Enterococcus faecalis* by a direct-current, cold atmospheric-pressure air plasma microjet. J. Biomed. Res. *24*, 264–269.

Topala, I., and Nagatsu, M. (2015). Capillary plasma jet: a low volume plasma source for life science applications. Appl. Phys. Lett. *106*, 054105.

WHO. (2003). Available from: <http://www.who.int/mediacentre/factsheets/2003/fs211/en/>

Zimmermann, J.L., Dumler, K., Shimizul, T., Morfill, G.E., Wolf, A., Boxhammer, V., Schlegel, J., Gansbacher, B., and Anton, M. (2011). Effects of cold atmospheric plasmas on adenoviruses in solution. J Phys. D: Appl. Phys. *44*, 505201.

Current Technology and Applications of Gas Plasma for Disinfection of Agricultural Products: Disinfection of Fungal Spores on *Citrus unshiu* by Atmospheric Pressure Dielectric Barrier Discharge

11

Yoshihito Yagyu and Akikazu Sakudo

Abstract

In the field of agricultural and food processes, the development of safe, high-quality disinfection methods that do not rely on chemical treatment is a promising approach for preventing food poisoning caused by pathogens such as fungi and bacteria. Recently, information has accumulated showing the potential of gas plasma as a novel food disinfection technology. In this chapter, to highlight recent advances in gas plasma technology for applications in food disinfection, we introduce our studies on the disinfection of *Penicillium digitatum* spores on *Citrus unshiu* by atmospheric pressure dielectric barrier discharge (APDBD). Future perspectives of this technology will also be discussed.

Introduction

Food security and food safety are crucial factors with regard to protection against food poisoning and plant disease in the food quality industry and in the management of agricultural production and food processing. With the exception of organic farming, agricultural chemicals such as pesticides and fungicides are usually applied to protect agricultural crops from hazardous injuries caused by disease and harmful insects. Postharvest diseases cause substantial postharvest losses, which is one of the major issues with agricultural crops. To avoid such losses, agricultural products are generally sterilized with substances exerting chemical and physical bactericidal effects, such as agricultural chemicals, chlorine bactericides, bactericidal gas, γ-ray irradiation, or thermal treatment. Application of agricultural chemicals for agricultural crops after harvest is strictly prohibited under Japanese regulations, although postharvest applications of other chemical-free methods are permitted. Therefore, it is a challenge to protect agricultural products from injury caused by pests and several fungi and viral bacterial diseases during storage or transportation. Furthermore, there is currently a general consumer concern of the residual agricultural chemicals in fresh fruits and vegetables. In actuality, most farmers, agricultural workers, and producers of the agriculture market tend to support pesticide-free or pesticide reduction production (Misra *et al.*, 1991; Fu *et al.*, 1999; Krystallis and Chryssohoidis, 2005; Dimitri and Dettmann, 2012).

There is always a risk of exposure of an agricultural product to a plant disease caused by harmful fungi or other pests before it reaches the consumer. For example, the green mould *Penicillium digitatum* and the blue mould *Penicillium italicum* cause important fungal diseases of all types of citrus. According to statistics released by the Ministry of Agriculture, Forestry, and Fisheries of Japan in 2011, approximately 100,000 tons of the satsuma mandarin *Citrus unshiu*, a common and popular species of Japanese citrus, were lost during postharvest, and the financial losses were estimated at around 20 billion yen per year.

To improve the safety and security of agricultural crops, we investigated an alternative method to the use of agrichemicals for prevention against fungal diseases using plasma disinfection. Low-temperature sterilization and disinfection techniques using plasma have been developed for medical applications (Hayashi *et al.*, 2006; Utsumi *et al.*, 2014; Iseki *et al.*, 2012), and the discovery of novel plasma applications for the life sciences has become one of the major topics in the field of plasma treatment of biological materials (Moisan *et al.*, 2001; Laroussi, 2005; Kitazaki *et al.*, 2012; Dubinov *et al.*, 2000). In this study, the effect of plasma disinfection using atmospheric pressure dielectric barrier discharge (APDBD) against green mould spores attached to citrus fruit was investigated (Yagyu *et al.*, 2015). We further discuss the possibility of the plasma disinfection technique for practical agricultural applications as an alternative method to the use of agrichemicals.

Direct plasma disinfection of green mould spores on citrus fruit by APDBD

APDBD as an atmospheric plasma source was generated by an AC high-voltage power unit (Logy Electric Co., LHV-10AC) and an electrode for dielectric barrier discharge (Fig. 11.1). The output waveform of the energy source had a sinusoidal wave shape at 10 kVpp with

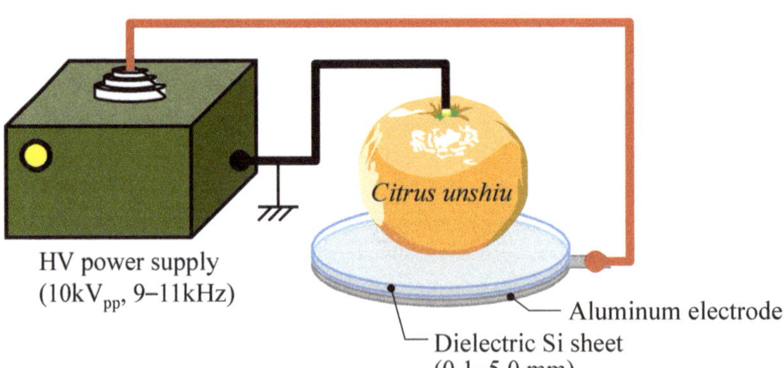

Figure 11.1 Experimental setup for direct plasma disinfection of green mould spores on the surface of citrus.

9–11 kHz. The APDBD electrode used in this study consisted of an aluminium plate covered by a silicon (Si) sheet with a diameter of 9.0 cm and a thickness ranging from 0.05 mm to 5 mm. The spore suspension of the green mould *P. digitatum* was prepared with 50 ppm dioctyl sulfosuccinate sodium salt base, and the density of mould spores was maintained at 1×10^8 cfu/ml *P. digitatum* spores. The suspension was directly applied to the surface of the satsuma mandarin *C. unshiu*, and then naturally dried in the atmosphere.

The irradiation period of APDBD varied from 0 to 5 s, and was generated on the surface of citrus so that the mould spores on the citrus were directly exposed to plasma in an open environment. Spores on the citrus were picked up with a wet swab tip (ST-25PBS, Elmex Ltd.), and the swab was vortexed for 10 s to release the spores from the swab tip into phosphate-buffered saline. The effect of APDBD on inactivation of the mould spores was estimated by using a sheet-type medium for yeast and mould (Sanita-kun, Chisso Corp.) (Teramura *et al.*, 2015). Mould spores were cultured in an incubator maintained at a constant temperature of 25°C, and then viable mould spores appeared on the medium sheet as red-coloured colonies after 48 to 72 h.

An aluminium plate covered by a dielectric silicon sheet was used as the APDBD electrode, and plasma was generated on the surface of the object by applying a power of 10 kV between the electrode and the object.

Condition of APDBD plasma generated on citrus

The typical condition of APDBD plasma is shown in Fig. 11.2. APDBD plasma of a brightly luminescent area was observed at a boundary surface between the electrode and object on the circumference of *C. unshiu* from the side view exposing the plasma (Fig. 11.2a). APDBD plasma was also generated on the surface of *C. unshiu*, and double concentric circles of the light emitted by the plasma were observed from the bottom view, as shown in Fig. 11.2(b). Furthermore, small and numerous bright spots were observed, which were assumed to be the oil glands of citrus.

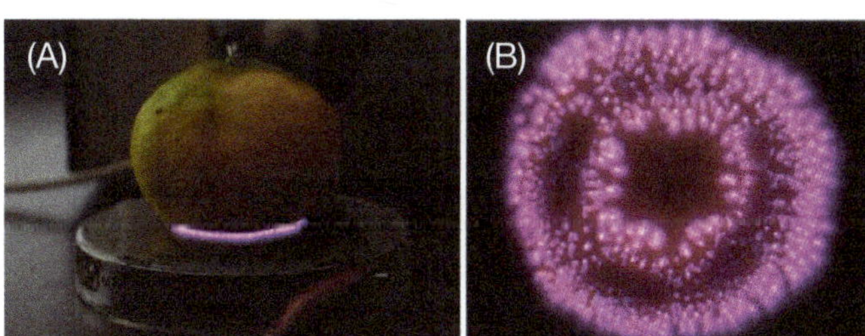

Figure 11.2 APDBD plasma generated on the surface of *Citrus unshiu*. (a) Side view with plasma irradiation. (b) Bottom view with plasma irradiation. Violet-coloured plasma was observed at a bright boundary surface between the APDBD electrode and *Citrus unshiu*. APDBD plasma was generated by two-fold concentric circles. Small and numerous bright spots were observed, which were assumed to be the oil glands of citrus.

Disinfecting effect of green mould spores on citrus by direct treatment of APDBD plasma

P. digitatum spores on the surface of *C. unshiu* were disinfected by APDBD plasma. The total viable mould spore count was clearly reduced depending on the irradiation period of APDBD plasma (Fig. 11.3). UV radiation and reactive oxygen species such as ozone, O radicals, and OH radicals contributed as inactivation factors in APDBD plasma. Previous studies showed that an atomic oxygen radical could effectively inactivate *P. digitatum* spores in atmospheric pressure plasma (Iseki *et al.*, 2010; Iseki *et al.*, 2011; Hashizume *et al.*, 2014). Furthermore, UV-C radiation in plasma was reported to be the dominant factor responsible for the inactivation of *P. digitatum* (Gündüz and Pazir, 2013).

Next, the effect of the thickness of the dielectric sheet used for inactivation of *P. digitatum* spores was examined. At a thickness of 0.3 mm, the number of germinating spores was reduced by approximately 1/10 in 1 s of irradiation and by approximately 1/100 in 3 s of irradiation. Furthermore, the disinfecting effect of the thin sheet (1.0 mm) was higher than that of the thick sheet (5.0 mm). On the other hand, there was no clear difference in disinfecting effects among sheets less than 1.0 mm thick. The potential difference between the aluminium electrode and Si sheet was measured, which tended to be gradually saturated as the thickness of the dielectric sheet increased, as shown in Fig. 11.4. Further, the energy consumption of APDBD increased with decreasing thickness of the Si sheet. The condition of APDBD plasma showed a tendency to change drastically depending on the thickness of the dielectric Si sheet when the thickness was less than 1.0 mm.

When the thickness of the Si sheet was less than 0.3 mm, approximately 1/10 of the spores were inactivated in 1 s of irradiation, and approximately 1/100 of the spores were inactivated in 3 s of irradiation of APDBD.

APDBD plasma showed a tendency to change drastically when the thickness of the silicon sheet was less than 1.0 mm.

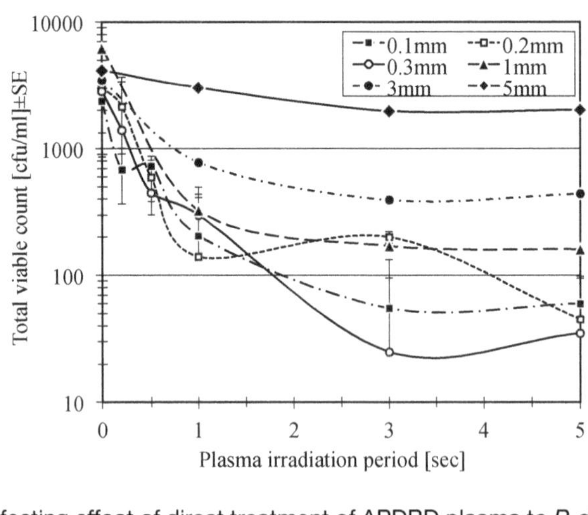

Figure 11.3 Disinfecting effect of direct treatment of APDBD plasma to *P. digitatum* spores on the surface of *Citrus unshiu* depending on the irradiation period and Si sheet thickness.

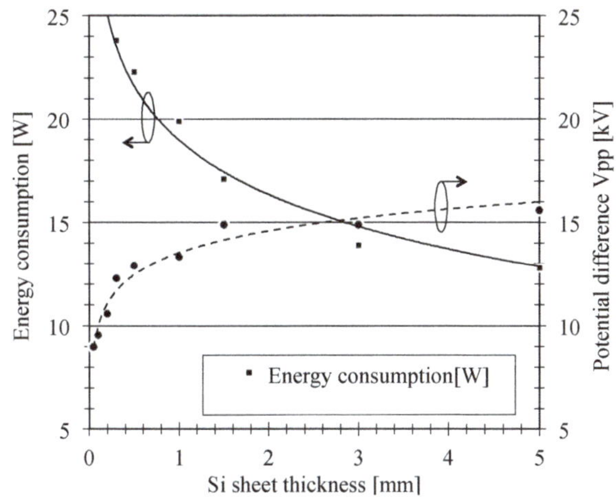

Figure 11.4 Thickness dependency of the dielectric Si sheet for the potential difference and energy consumption of APDBD.

Relationship between surface temperature and thermal injury of citrus

A thermography camera (VarioCAM® High resolution basic 384, Jenoptik Ltd.) was used to measure the surface temperature of citrus during APDBD plasma exposure, and the surface condition of citrus was visually observed at each irradiation period. As shown in the thermographic images presented in Fig. 11.5, the surface temperature partially increased immediately after plasma irradiation. A significant temperature increase was noted at a boundary surface between the electrode and object on the circumference of *C. unshiu*.

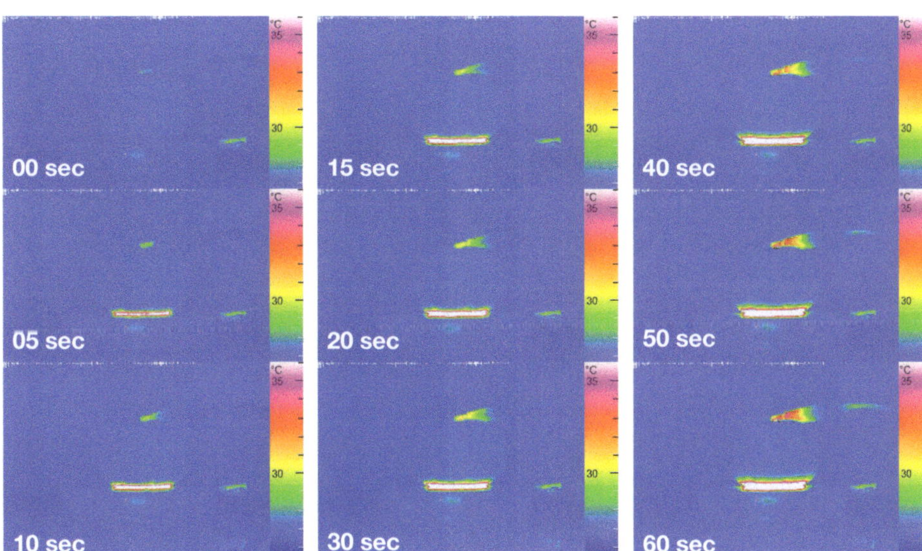

Figure 11.5 Thermographic images of *Citrus unshiu* surface.

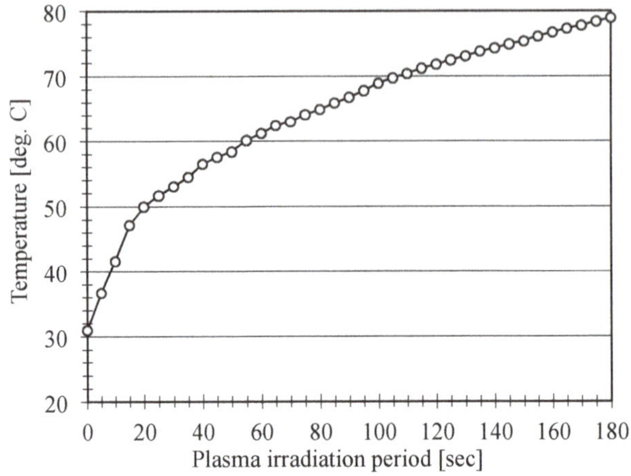

Figure 11.6 Relationship between the surface temperature of *Citrus unshiu* and exposure period of APDBD plasma.

Fig. 11.6 shows the change in surface temperature of citrus depending on the exposure period of APDBD plasma. The surface temperature of citrus treated with APDBD increased rapidly and then gradually saturated with increasing irradiation time. The surface temperature also increased depending on the thickness of the Si sheet, with a higher temperature observed with the 1.0-mm-thick sheet compared to the 0.5-mm-thick sheet. The energy consumption of APDBD was also strongly dependent on the thickness of the Si sheet.

Fig. 11.7 indicates the visual condition of citrus from 0 s of irradiation, as a control, to 3 min of irradiation, when APDBD plasma was generated with the 0.5-mm-thick Si sheet. Although the citrus surface was not injured following plasma irradiation for 5 s to 30 s, the

Figure 11.7 Thermal injury of *Citrus unshiu* induced by APDBD plasma disinfection from 0 s to 3 min.

surface of *C. unshiu* turned brown after 1 min of irradiation. Therefore, under the condition of less than approximately 65°C with 60 s of irradiation time, the surface of citrus did not incur any damage. However, the citrus surface was obviously injured by heat generated from an electrode when the temperature increased over 65°C with increasing treatment time. However, the negative effect of thermal damage of the citrus surface induced by plasma treatment would be negligible from a practical standpoint, because APDBD plasma could reduce the viable count of *P. digitatum* spores on *C. unshiu* to about 1/100 with only 3 s of treatment (Fig. 11.3).

The surface temperature partially increased at a boundary surface between the electrode and the circumference of *Citrus unshiu* immediately after plasma irradiation, and a significant temperature rise was recognized.

The surface temperature of citrus treated with APDBD increased rapidly and then gradually saturated with increasing irradiation time.

The citrus surface was not injured until 30 s of plasma irradiation. The surface of *Citrus unshiu* turned brown at irradiation times longer than 1 min. The damage occurred by heat from the electrode when the temperature increased over 65°C with increasing treatment time.

Typical optical emission spectrum from APDBD on the citrus surface

Fig. 11.8 shows a typical optical emission spectrum in the range of 250–800 nm from APDBD plasma, which was generated in an open environment by the device illustrated in Fig. 11.1. The spectrum was measured on a spectrometer (Hamamatsu, PMA-C8808), and the optical emission from the N_2 second positive band showed high intensity. Other species were also observed, such as the N_2 first positive band at around 400 nm and the O radical at 777 nm, although these were very weak compared to the N_2 second positive band (Fig. 11.9). Furthermore, UV emissions, such as UV-A (320–400 nm), UV-B (290–320 nm), and UV-C (200–290 nm), were confirmed from the plasma generated on the citrus surface,

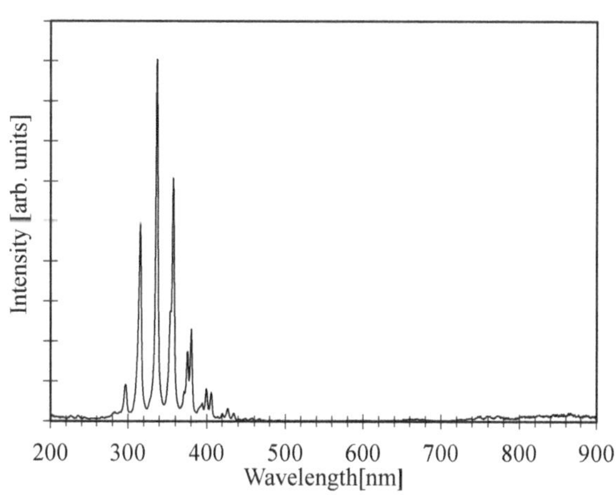

Figure 11.8 Typical optical emission spectrum of air plasma.

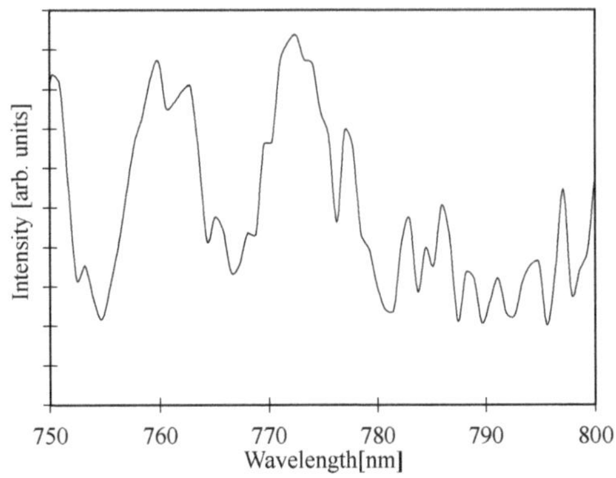

Figure 11.9 Typical optical emission spectrum of air plasma between 750 and 800 nm. Oxygen radicals at 777 nm were observed in the spectrum from APDBD plasma generated in an open environment.

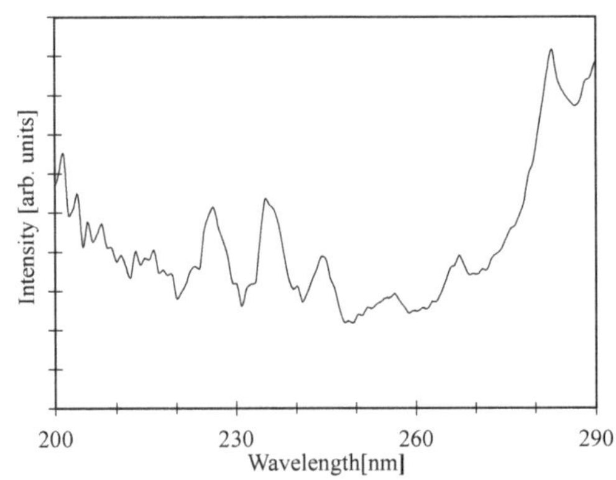

Figure 11.10 Typical optical emission spectrum of air plasma between 200 and 290 nm. UV radiation categorized as UV-C (200–290 nm) was observed in the spectrum from APDBA plasma generated in an open environment.

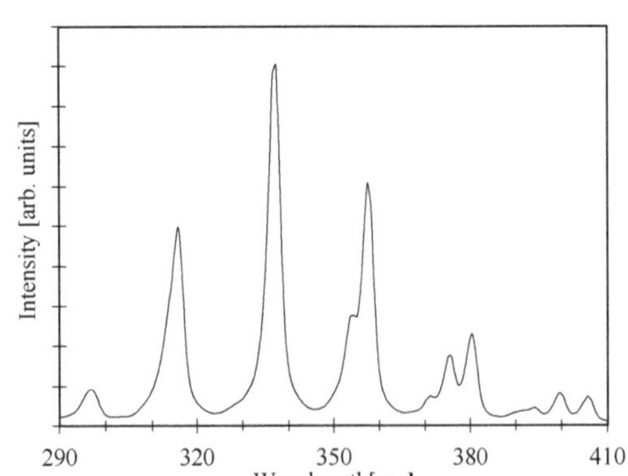

Figure 11.11 Typical optical emission spectrum of air plasma between 290 and 410 nm. UV radiation categorized as UV-B (290–320 nm) and UV-A (320–400 nm) was observed in the spectrum from APDBD plasma generated in an open environment.

as indicated in Fig. 11.10 and Fig. 11.11. Therefore, the dominant inactivation factors of APDBD were considered to be UV radiation and active oxygen species such as ozone, O radicals, and OH radicals.

Oxygen radicals at 777 nm were observed in the spectrum from APDBD plasma generated in an open environment.

UV radiation categorized as UV-C (200–290 nm) was observed in the spectrum from APDBD plasma generated in an open environment.

UV radiation categorized as UV-B (290–320 nm) and UV-A (320–400 nm) was observed in the spectrum from APDBD plasma generated in an open environment.

Conclusions and future perspectives

Disinfection of *Citrus unshiu* using direct application of APDBD plasma for inactivating *P. digitatum* spores was investigated in this study. APDBD plasma irradiation showed good ability to sufficiently inactivate mould spores in only a few seconds with no damage to the surface of citrus. The surface temperature of citrus depended on the exposure period of APDBD plasma. The citrus surface was thermally discoloured to a brownish colour when the temperature was higher than 65°C. However, the damage caused by high temperature would not be a practical problem because mould spores were effectively treated in only a few seconds before a high temperature was reached.

The novel technique of plasma irradiation of fresh food and processed foods such as vegetables, crops, and fruits shows potential to be widely adopted in agricultural applications. Furthermore, novel plasma applications have been developed for clinical medicine, and the positive effects of plasma on biological materials such as cells and the living state have been verified in recent studies. Practical applications of plasma have also been developed including our technology. It is expected that plasma disinfection will be widely realized as an alternative method to agrichemicals application in the near future.

Acknowledgements

This study was supported in part by the Science and Technology Research Promotion Program for Agriculture, Forestry, Fisheries and Food Industry.

References

Dimitri, C., and Dettmann, R.L. (2012). Organic food consumers: What do we really know about them? British Food J. *114*, 1157–1183.

Dubinov, A.E., Lazarenko, E.M., and Selemir, V.D. (2000). Effect of glow discharge air plasma on grain crops seed. IEEE Tran. Plasma Sci. *28*, 180–183.

Fu, T.T., Liu, J.T., and Hammitt, J.K. (1999). Consumer willingness to pay for low pesticide fresh produce in Taiwan. J. Agr. Econ. *50*, 220–233.

Gündüz, G.T., and Pazir, F. (2013). Inactivation of *Penicillium digitatum* and *Penicillium italicum* under in vitro and in vivo conditions by using UV-C light. J. Food Prot. *76*, 1761–1766.

Hashizume, H., Ohta, T., Takeda, K., Ishikawa, K., Hori, M., and Ito, M. (2014). Oxidation mechanism of *Penicillium digitatum* spores through neutral oxygen radicals. Jpn. J. Appl. Phys. *53*, 010209:1–6.

Hayashi, N., Guan, W., Tsutsui, S., Tomari, T., and Hanada, Y. (2006). Sterilization of medical equipment using radicals produced by oxygen/water vapour RF plasma. Jpn. J. Appl. Phys. *45*, 8358–8363.

Iseki, S., Hashizume, H., Jia, F., Takeda, K., Ishikawa, K., Ohta, T., Ito, M., and Hori, M. (2011). Inactivation of *Penicillium digitatum* spores by a high-density ground-state atomic oxygen-radical source employing an atmospheric-pressure plasma. Appl. Phys. Express 4, 116201:1–3.

Iseki, S., Nakamura, K., Hayashi, M., Tanaka, H., Kondo, H., Kajiyama, H., Kano, H., Kikkawa, F., and Hori, M. (2012). Selective killing of ovarian cancer cells through induction of apoptosis by nonequilibrium atmospheric pressure plasma. Appl. Phys. Lett. *100*, 113702:1–4.

Iseki, S., Ohta, T., Aomatsu, A., Ito, M., Kano, H., Higashijima, Y., and Hori, M. (2010). Rapid inactivation of *Penicillium digitatum* spores using high-density nonequilibrium atmospheric pressure plasma. Appl. Phys. Lett. *96*, 153704:1–3.

Kitazaki, S., Koga, K., Shiratani, M., and Hayashi, N. (2012). Growth enhancement of radish sprouts induced by low pressure O_2 radio frequency discharge plasma irradiation. Jpn. J. Appl. Phys. *51*, 01AE01:1–4.

Krystallis, A., and Chryssohoidis, G. (2005). Consumers' willingness to pay for organic food: factors that affect it and variation per organic product type. G. British Food J. *107*, 320–343.

Laroussi, M. (2005). Low temperature plasma-based sterilization: overview and state-of-the-art. Plasma Process. Polym. *2*, 391–400.

Misra, S.K., Huang, C.L., and Ott, S.L. (1991). Consumer willingness to pay for pesticide-free fresh produce. Western J. Agr. Econ. *16*, 218–227.

Moisan, M., Barbeau, J., Moreau, S., Pelletier, J., Tabrizian, M., and Yahia, L'H. (2001). Low-temperature sterilization using gas plasmas: a review of the experiments and an analysis of the inactivation. Mechanisms Int. J. Pharm. *226*, 1–21.

Teramura, H., Ushiyama, M., and Ogihara, H. (2015). Evaluation of a novel dry sheet culture method (Sanita-kunR) for rapid enumeration of yeasts and molds in foods. J. Microbiol. Meth. *109*, 16–19.

Utsumi, F., Kajiyama, H., Nakamura, K., Tanaka, H., Hori, M., and Kikkawa, F. (2014). Selective cytotoxicity of indirect nonequilibrium atmospheric pressure plasma against ovarian clear-cell carcinoma. Springerplus, 3, 398:1–9.

Yagyu, Y., Hatayama, Y., Hayashi, N., Mishima, T., Nishioka, T., Sakudo, A., Ihara, T., Ohshima, T., Kawasaki, H., and Suda, Y. (2015). Direct plasma disinfection of green mold spore on citrus by atmospheric pressure dielectric barrier discharge for agricultural applications. Trans. Mat. Res. Soc. Japan in press.

Current Progress in Seed Disinfection by Gas Plasma: Disinfection of Seed-borne Fungi and Bacteria by Plasma with Alternating Current High-voltage Discharge

12

Terumi Nishioka, Tomoko Mishima, Yoichi Toyokawa, Tatsuya Misawa and Akikazu Sakudo

Abstract

Seed-borne pathogens are one of the most important causative agents of plant diseases and prevent healthy growth, resulting in the reduction of marketable and profitable crops. Currently, treatments with hot water and air have been used for disinfection of seeds; however, these methods are time-consuming and sometimes damage seeds. Recent studies have shown the potential of gas plasma technology for disinfection of seeds. As short plasma treatments have achieved successful disinfection of seed-borne fungi and bacteria without seed damage, this innovative technology can facilitate rapid and safe disinfection of seeds. This chapter will outline our recent studies on disinfection of seed-borne fungi and bacteria by plasma with alternating current high-voltage discharge and discuss the future perspectives for seed disinfection made possible by plasma technology.

Introduction

Seeds infested with plant pathogens may reduce seed germination or vitality owing to the development of diseases. The resulting decrease in the seedling population leads great loss in crop yield. Additionally, these seeds carrying plant disease agents may introduce diseases into new, pathogen-free fields (Kulik, 1995).

Of the various kinds of seed-borne plant pathogens, fungi are responsible for the greatest number of plant diseases. Effective control measures include exclusion of pathogens through seed certification and seed treatment (Neergaard, 1977). The use of seed dressing or dipping seeds in fungicide is very effective and commonly used. However, in cases where a novel disease occurs or originates from imported seeds, suitable chemicals might not be domestically available. Seed-borne pathogenic bacteria represent the second largest group of plant pathogens (Kulik, 1995), but compared to fungicides, the variety of bactericides available in Japan is very limited. Moreover, in recent years, there has been increasing public demand for ecofriendly alternatives to fungicides and bactericides.

There are two commonly used, ecofriendly methods for seed disinfection: hot water and hot air treatments. Hot water treatments are generally completed in 10–30 min at around 50°C; however, the water temperature and treatment time must be carefully controlled to inactivate the pathogens without damaging the seeds (Babadoost, 1992). The hot air treatments do not involve a drying process after treatment, and are advantageous in comparison to hot water treatments. But more time (several days) is required to inactivate the pathogens.

Plasma treatment has attracted a lot of attention as an ecofriendly method for inactivating microorganisms, and the application of plasma sterilization to seed disinfection has been studied. Selcuk *et al.* (2008) reported that their self-designed low-pressure cold plasma system could significantly reduce surface fungal contamination of seed storage fungi, such as *Aspergillus* spp. and *Penicillium* spp., on grains and legumes. They achieved a 3-log reduction of *Penicillium* spp. on wheat within 15 min. The plasma was generated at running pressure 0.07 kPa with air gasses or SF_6 as the discharge gas in their system. Filatova *et al.* (2012) reported that low-pressure radio-frequency air plasma treatment for 5–15 min reduced the level of seed infection with pathogenic fungi such as *Fusarium* spp., *Alternaria* spp., and *Stemphylium* spp. These studies suggest that plasma treatment can inactivate pathogens on seeds more rapidly than hot air treatment and without wetting the seeds. However, the number of reports concerning seed disinfection by plasma is small.

Hence, in this chapter, we will discuss our studies on the disinfection of seeds contaminated with pathogenic fungi and bacteria.

Disinfecting effect of plasma treatment on seed-borne fungi and bacteria

Electric discharge under low gas pressure conditions can generate a large volume of uniform plasma in comparison with discharge at atmospheric pressure conditions, and low-pressure and low-temperature plasma might be better suited for the future application of gas plasma to seed disinfection (Nishioka *et al.*, 2014). We evaluated the effect of low-pressure plasma on the fungal and bacterial inactivation. The plasma apparatus is shown in Fig. 12.1 (Nishioka *et al.*, 2014). The plasma was generated by low frequency AC discharge at a low gas pressure. Two insulated aluminium electrodes were installed to a level in the glass chamber. The pressure in the chamber was controlled by the vacuum pumping and the discharge gas injection. Argon gas was used as discharge gas. When the flow rate of argon gas was 0.5 L/min, the pressure in the chamber was sustained at 10.7 kPa. A high voltage was applied to the electrodes by the low frequency AC power supply, and low-pressure and low-temperature argon plasma was generated between the electrodes. The peak-to-peak voltage and frequency of power supply were 5.5 kV and 10 kHz, respectively. The seeds were placed on the mesh sheet between electrodes and exposed to the argon plasma.

As a model of seed-borne fungi, *Rhizoctonia solani*, which causes damping-off, was inoculated on brassicaceous seeds (*Brassica campestris* var. *amplexicaulis*). This fungus can be transmitted by seeds of solanaceous and brassicaceous crops and Japanese hornwort (*Cryptotaenia japonica* Hassk) (Neergaard, 1977; Fujita *et al.*, 2005). Japanese hornwort is usually grown by soilless culture in Japan and once the damping-off appears, no available chemical agents can effectively control the disease. Nishioka *et al.* (2014) reported that plasma treatment for 5 min reduced the fungal survival rate to 3.3%, while the inoculated *R. solani* was found to be alive on 83% of the seeds before treatment. The inactivation effect

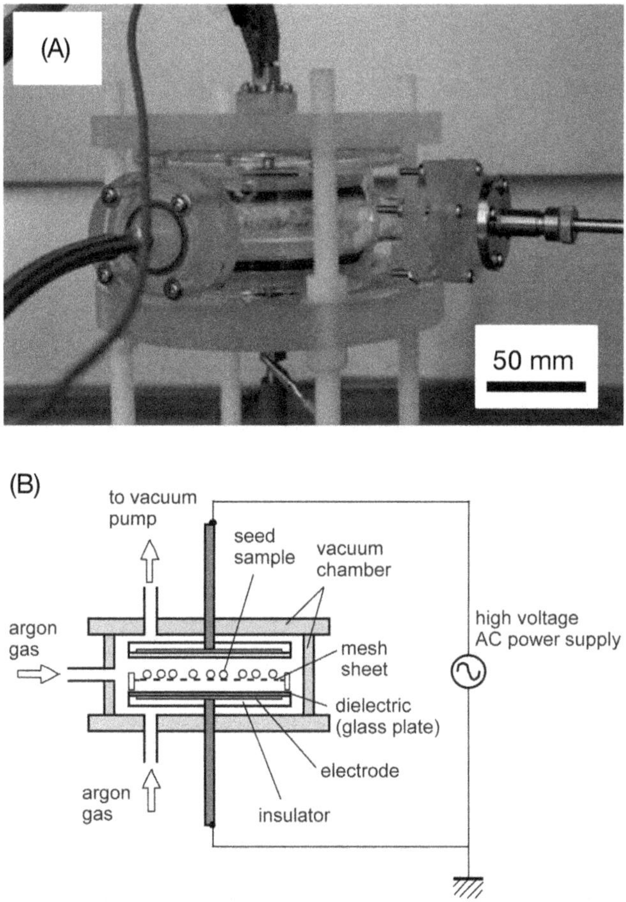

Figure 12.1 Low-pressure plasma apparatus (a) and its schematic diagram (b). Plasma was generated by an AC high-voltage discharge and argon was used. Seeds were set on a mesh sheet in the plasma apparatus. Reproduced from Nishioka *et al.* (2014) with permission from the Society for Antibacterial and Antifungal Agents, Japan.

on *R. solani* was dependent on the plasma treatment time. Treatment for 40 min achieved complete inactivation of *R. solani* on the seeds (Fig. 12.2).

Black rot caused by *Xanthomonas campestris* pv. *campestris* is considered the most important disease of crucifers worldwide, attacking all cultivated brassicas, radishes, and numerous cruciferous weeds (Williams, 1980). Its infested seeds are a primary source of the disease (Shiomi, 1992). Hence, *X. campestris* was used as a model of seed-borne bacteria in this study. The inactivation effect on *X. campestris* inoculated on brassicaceous seeds was also dependent on the plasma treatment time (unpublished results). The viable cell number was reduced to below 0.1% after plasma treatment for a few minutes.

We also evaluated the inactivation of *Magnaporthe oryzae,* a pathogenic fungi, by the atmospheric pressure plasma (Misawa *et al.*, 2013; Nishioka *et al.*, 2013). *M. oryzae* causes rice blast disease, which is one of the most important diseases in the rice-producing regions of the world (Agrios, 2005). More than 5-min plasma treatment with an AC high-voltage

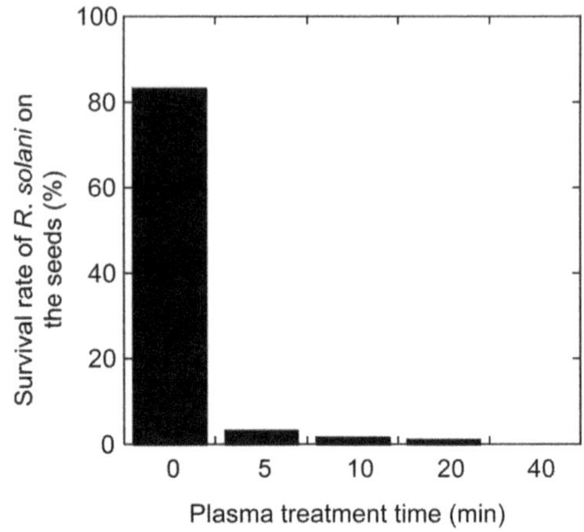

Figure 12.2 Inactivation of *Rhizoctonia solani* inoculated on seeds with low-pressure plasma. Plasma treatment was performed using the plasma apparatus shown in Figure 12.1. The voltage and frequency applied to electrodes were 5.5kV and 10kHz, respectively. The argon gas flow rate was 0.5 L/min. The running pressure in the chamber was 10.7 kPa. *Rhizoctonia solani* was inoculated on *Brassica campestris* var. *amplexicaulis* seeds. The *R. solani* survival rate on seeds after plasma treatment was observed. Reproduced from Nishioka *et al.* (2014) with permission from the Society for Antibacterial and Antifungal Agents, Japan.

discharge of 10kV at 10kHz and a flow rate of 3 l/min argon, could inhibit the conidial germination of *M. oryzae* (Fig. 12.3).

Since the early 1970s, there has been a global emergence of fungicide-resistant strains (Ishii, 2006), and many of these strains are also seed-borne pathogens including *M. oryzae.* Seed disinfection by plasma could be an important strategy for disinfection, which is difficult to achieve using traditional fungicides or bactericides, against such chemicals-resistant pathogens. Further investigations are needed for the application to various kinds of seeds, since the surface structure and hull composition of seeds also have an impact on the effectiveness of the plasma (Selcuk *et al.*, 2008).

Heat generated during plasma treatment and its influence on seed germination

Several factors such as heat, ultraviolet (UV) radiation, and oxidative stress are generally produced during the plasma generation process (Maeda *et al.*, 2015). The generated dry heat could inactivate plant pathogens on seeds, while simultaneously damaging the seeds. In general, hot air treatments are carried out at about 70–80°C, and it takes 2–5 days to inactivate pathogens. The temperature in our low-pressure plasma apparatus was 50–60°C after plasma treatment for 5 min. It gradually rose and finally reached a temperature higher than 70°C after 40 min. We evaluated the germination rate as an indicator of seed quality (Nishioka *et al.*, 2014). There was no significant difference in the germination rate of *B. campestris* var. *amplexicaulis* seeds between the untreated and each plasma-treated group ($P > 0.05$; Table 12.1). These facts suggest that the low-pressure plasma system does not induce significant thermal damage on the seeds.

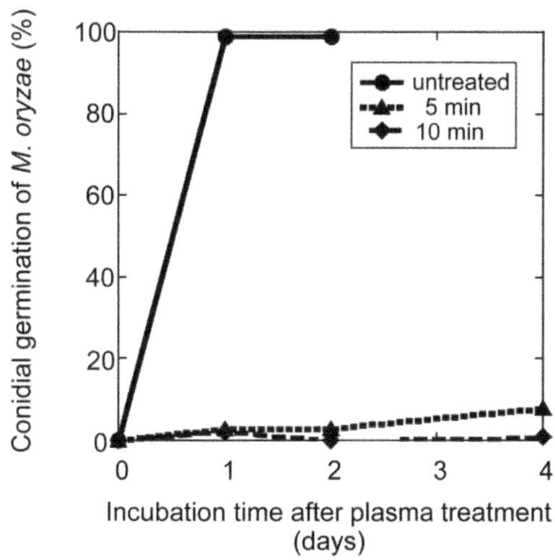

Figure 12.3 Inhibition of *Magnaporthe oryzae* conidial germination with atmospheric-pressure plasma. The plasma was generated by AC high-voltage discharge with argon. The voltage and frequency applied to electrodes were 10 kV and 10 kHz, respectively. The argon gas flow rate was 3 L/min. The distance from the plasma source to the sample surface was 3 cm. The 0.8-cm water agar-cored disk dispensed with *M. oryzae* conidia, approximately 5 × 10^2 per disc, were treated with plasma. After each treatment, the *M. oryzae* disc was incubated on a fresh potato dextrose agar plate at 25°C to evaluate the conidial germination rate. Reproduced from Nishioka *et al.* (2013) with permission from the Asian Food Safety and Security.

Table 12.1 The germination rate of *Brassica campestris* var. *amplexicaulis* seeds (%) after low-pressure plasma treatment (Nishioka *et al.*, 2014)

	Post-treatment time	
Treatment time	2 days	6 days
untreated	95.6	96.7
5 min	100	100
10 min	93.3	96.7
20 min	90.0	94.4
40 min	90.0	95.0

The voltage and frequency applied to electrodes were 5.5kV and 10 kHz, respectively. The argon gas flow rate was 0.5 L/min. The running pressure in the chamber was 10.7 kPa. No significant differences in seed germination were found among untreated and plasma treated groups both after 2 days and 6 days' incubation ($P > 0.05$).

In our low-pressure plasma system, dry heat might not be an important factor for fungal and bacterial inactivation. The dry heat equivalent to that generated during plasma treatment could not significantly reduce the cell number of *X. campestris* on the seeds (unpublished results). Hot air treatments for seed disinfection of *X. campestris* at 75°C take 5–7 days (Shiomi, 1992), while disinfection of *R. solani* on Japanese hornwort seeds takes 3 days at 77°C (Kawaradani *et al.*, 2009). Thus, some factors other than dry heat might be associated with inactivation of the pathogens.

An air plasma discharge produces oxygen atoms, ozone, OH-radicals, N-radicals, plasma electrons, and so on (Deng *et al.*, 2007). Such species are very strong oxidizers and are thought to play important roles in damaging microorganisms. Gaunt *et al.* (2006) described that ozone produced in the gas discharge is one of the most significant reactive species that causes oxidative damage. Its density depends on the plasma apparatus and plasma generation conditions. Iseki *et al.* (2010) reported that their high-density non-equilibrium atmospheric pressure plasma employing argon gas successfully inactivated *Penicillium digitatum* spores. The ozone density reached 2 to 8 ppm during plasma generation. They confirmed that the contribution of ozone to inactivation of *P. digitatum* spores was small. In our low-pressure plasma, a very small amount of ozone was generated, but its density was not considered to be high enough to inactivate pathogens such as *R. solani* and *X. campestris* (unpublished results).

In recent years, surface modification of seeds with plasma discharge has attracted a lot of attention as a method for improving seed germination rate and the speed. Positive effects of plasma treatment on germination of various agricultural and horticulture plants have been previously demonstrated (Dhayal *et al.*, 2006; Filatova *et al.*, 2012; Mitra *et al.*, 2014; Šerá *et al.*, 2013). On the other hand, Šerá *et al.* (2012) compared four plasma treatments and confirmed that stimulation or inhibition of germination and early growth of seed are strongly connected not only with plant species and duration of plasma exposure but also with the type of plasma apparatus. Barmashenko *et al.* (2012) demonstrated the modification of the wettability characteristics of seeds by cold radiofrequency air plasma treatment, and they stated that it was reasonable to relate the change of wettability of seeds to the oxidation of their surfaces under plasma treatment.

Seed disinfection by plasma for food safety

Many foodborne illness outbreaks associated with the consumption of raw sprouts have occurred in the past few decades worldwide. In most of the cases, *Escherichia coli* or *Salmonella* spp., the causative organisms, is considered to originate from sprouted seeds, and the high temperature and humidity during seed germination are suitable for the rapid growth of pathogens (Castro-Rosas and Escartín, 2000; Feng *et al.*, 2007; Saroj *et al.*, 2007; Taormina and Beuchat, 1999). Accordingly, the decontamination of seeds prior to germination is important for the safety of sprouted seeds (Nei *et al.*, 2013).

Chemical treatments with chlorine are currently used to disinfect sprout seeds. However, high levels of chlorine discharged into wastewater treatment facilities present a large environmental burden (Hu *et al.*, 2004), and such treatments also have an impact on human health. In some European countries the use of chlorine for the ready- to use food products is prohibited (Ongeng *et al.*, 2006; Rico *et al.*, 2007). The need for other disinfection and sterilization technologies to control infection and disease is increasing.

Recently, there have been numerous studies evaluating the inactivation of food-borne pathogens such as *E. coli* O157:H7, *Salmonella* spp., and *Listeria monocytogenes*, and several reports show that plasma treatment can be used for reducing microbial populations on product surfaces (Critzer *et al.*, 2007; Fernández *et al.*, 2013; Niemira and Sites, 2008). Atmospheric pressure air plasma, which is based on a microwave driven discharge for 15 min, achieved an approximate 6-log CFU inactivation of *Bacillus atrophaeus* endospores, which are much more resistant to sterilization methods than vegetative bacteria, on various seeds (Schnabel *et al.*, 2012). Furthermore, the most recent study has evaluated the inactivation of natural microbiota attached to the chickpea seed surface by atmospheric pressure plasma (Mitra *et al.*, 2014). Although naturally occurring microorganisms can be attached on the seed surface, even in the narrow gaps, for a long time in the form of microcolonies or biofilms and are therefore harder to inactivate (Izquier and Gómez-López, 2011), it was shown that the plasma treatment for 2 and 5 min could achieve 1 and 2 log reduction, respectively.

Conclusion and further perspectives

In principle, plasma treatment may be primarily used to decrease microbial contaminants on surfaces (Schlüter *et al.*, 2013). Pathogens contaminate seeds during a number of processes including seed production, seed harvest, transport, storage, and via several pathways such as mixing with other infected seeds and unsanitary water. Gas plasma sterilization can be performed at low temperatures so as not to damage seeds, and with short processing times, without wetting seeds or using chemicals. Gas plasma may offer an alternative to current seed disinfection methods (Selcuk *et al.*, 2008).

Agricultural products may be exposed to several pathogens, which include seed-borne and soil-borne pathogens, that cause several food-borne illness, before being delivered to the consumer. Microbial hazard is one of the most important issues in food industry (Fig. 12.4). It is essential to control the probable pathogens at every step from farm to consumer. Numerous past studies have described that plasma treatment can inactivate various kinds

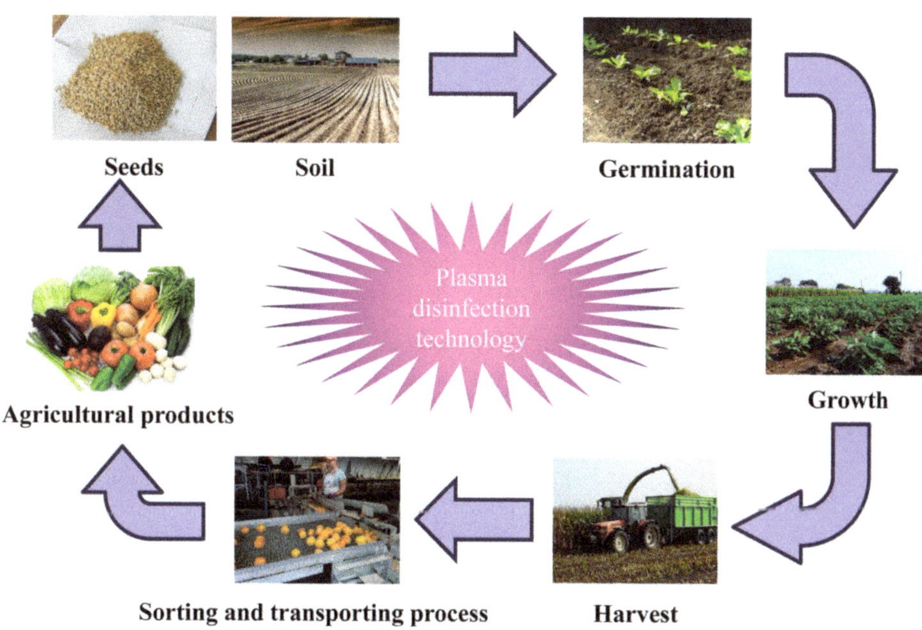

Figure 12.4 Future perspectives of plasma disinfection technology in agricultural field. Microbial hazards are one of the most important issues in food industry. Before delivering agricultural products, such as fruits and vegetables, plants are prone to contamination from water, soil, fertilizer, dusts, insects, animal faeces, and field workers during pre-harvest and harvest, in addition to transport, packaging and other food processing. Furthermore, consumers of contaminated fruits and vegetables might be exposed to a higher risk of food-borne illnesses, which may be caused by seed-borne and soil-borne pathogens.

of bacteria and fungi including several plant pathogens and human pathogens. To minimize the risk of disease, gas plasma can be used as an innovative disinfection technology to treat infested seeds before they grow into crops and processed as agricultural products.

Acknowledgement

This study was supported by the Programme for Promotion of Basic and Applied Researches for Innovations in Bio-oriented Industry and Science and technology research promotion program for agriculture, forestry, fisheries and food industry.

References

Agrios, G.N. (2005). Rice Blast Disease. In Plant Pathology fifth edition (Burlington, USA: Academic Press), pp.463.

Babadoost, M. (1992). Vegetable seed treatment. University of Illinois Extension: College of Agricultural, Consumer and Environmental Sciences. RPD No. 915. March 1992.

Barmashenko, E., Grynyov, R., Barmashenko, Y., and Drori, E. (2012). Cold radiofrequency plasma treatment modifies wettability and germination speed of plant seeds. Scientific Reports 2, 741.

Castro-Rosas, J., and Escartín, E.F. (2000). Survival and growth of *Vibrio cholerae* O1, *Salmonella typhi*, and *Escherichia coli* O157:H7 in alfalfa sprouts. J. Food Sci. *65*, 162–165.

Critzer, F.J., Kelly-Wintenberg, K., South, S.L., and Golden, D.A. (2007). Atmospheric plasma inactivation of foodborne pathogens on fresh produce surfaces. J. Food Prot. *70*, 2290–2296.

Deng, S., Ruan, R., Mok, C.K., Huang, G., Lin, X., and Chen, P. (2007). Inactivation of *Escherichia coli* on almonds using nonthermal plasma. J. Food Sci. 72, M62-M66.

Dhayal, M., Lee, S.-Y., and Park, S.-U. (2006). Using low-pressure plasma for *Carthamus tinctorium* L. seed surface modification. Vacuum *80*, 499–506.

Feng, G., Churey, J.J., and Worobo, R.W. (2007). Thermal inactivation of *Salmonella* and *Escherichia coli* O157:H7 on alfalfa seeds. J. Food Prot. *70*, 1698–1703.

Fernández, A., Noriega, E., and Thompson, A. (2013). Inactivation of *Salmonella enterica* serovar Typhimurium on fresh produce by cold atmospheric gas plasma technology. Food Microbiol. 33, 24–29.

Filatova, I., Azharonok, V., Shik, A., Antoniuk, A., and Terletskaya, N. (2012). Fungicidal effects of plasma and radio-wave pre-treatments on seeds of grain crops and legumes. In Plasma for bio-decontamination, medicine and food security (Nato Science for Peace and Security Series A: Chemistry and Biology), Z. Machala, K. Hensel and Y. Akishev, eds. (Heidelberg, Germany: Springer), pp. 469–479.

Fujita, T., Takimoto, M., Miyake, N., and Narita, S. (2005). Investigation of seed-borne diseases in Japanese hornwort. Ann. Rept. Kansai Pl. Prot. 47, 109–111.

Gaunt, L.F., Beggs, C.B., and Georghiou, G.E. (2006). Bactericidal action of the reactive species produced by gas-discharge nonthermal plasma at atmospheric pressure: a review. IEEE Trans. Plasma Sci. *34*, 1257–1269.

Hu, H., Churey, J.J., and Worobo, R.W. (2004). Heat treatments to enhance the safety of mung bean seeds. J. Food Prot. *67*, 1257–1260.

Iseki, S., Ohta, T., Aomatsu, A., Ito, M., Kano, H., Higashijima, Y., and Hori, M. (2010). Rapid inactivation of *Penicillium digitatum* spores using high-density nonequilibrium atmospheric pressure plasma. Appl. Phys. Lett. *96*, 153704.

Ishii, H. (2006). Impact of fungicide resistance in plant pathogens on crop disease control and agricultural environment. JARQ *40*(3), 205–211.

Izquier, A., and Gómez-López, V.M. (2011). Modeling the pulsed light inactivation of microorganisms naturally occurring on vegetable substrates. Food Microbiol. *28*, 1170–1174.

Kawaradani, M., Uryu, E., Nishida, A., Negoro, J., Okada, K., Nakasone, W., and Uchiyama, T. (2009). Seed heat treatment to control damping-off of Japanese hornwort. Bull. Res. Inst. Env. Agr. Fish Osaka, *2*, 25–28.

Kulik, M.M. (1995). Seed quality and microorganisms (Chapter 5). In Seed Quality: basic mechanisms and agricultural implications, A.S. Basra, ed. (New York, USA: Food Products Press), pp.153–171.

Maeda, K., Toyokawa, Y., Shimizu, N., Imanishi, Y., and Sakudo, A. (2015). Inactivation of *Salmonella* by nitrogen gas plasma generated by a static induction thyristor as a pulsed power supply. Food Control 52, 54–59.

Misawa, T., Ayabe, T., Yamamoto, K., Akiyoshi, Y., and Hayashi, N. (2013). Property of atmospheric plasma generated by water-dielectric multi-layer electrode. Front. Appl. Plasma Technol. *6*, 1–5.

Mitra, A., Li, Y.-F., Klämpfl, T.G., Shimizu, T., Jeon, J., Morfill, G.E., and Zimmermann, J.L. (2014). Inactivation of surface-borne microorganisms and increased germination of seed specimen by cold atmospheric plasma. Food Bioprocess Technol. *7*, 645–653.

Neergaard, P. (1977). Seed pathology. (London, UK: Macmillan Press Ltd.).

Nei, D., Bari, Md. L., and Enomoto, K. (2013). Validation of hot water and chlorine treatments to inactivate pathogens inoculated on mung bean seeds: Influence of the seed production area. Food Control *32*, 186–189.

Niemira, B.A., and Sites, J. (2008). Cold plasma inactivates *Salmonella* Stanley and *Escherichia coli* O157:H7 inoculated on Golden Delicious apples. J. Food Prot. *71*, 1357–1365.

Nishioka, T., Takai, Y., Kawaradani, M., Okada, K., Tanimoto, H., Misawa, T., and Kusakari, S. (2014). Seed disinfection effect of atmospheric pressure plasma and low pressure plasma on *Rhizoctonia solani*. Biocont Sci. *19*(2), 99–102.

Nishioka, T., Takai, Y., Kusakari, S., Kawaradani, M., Okada, K., Tanimoto, H., and Misawa, T. (2013). Inactivation of seed-borne pathogens using atmospheric gas plasma. Proceedings of First AFSSA Conference 26–28.

Ongeng, D., Devlieghere, F., Debevere, J., Coosemans, J., and Ryckeboer, J. (2006). The efficacy of electrolysed oxidizing water for inactivating spoilage microorganisms in process water and on minimally processed vegetables. Int. J. Food Microbiol. *109*, 187–197.

Rico, D., Martín-Diana, A.B., Barat, J.M., and Barry-Ryan, C. (2007). Extending and measuring the quality of fresh-cut fruit and vegetables: a review. Trends Food Sci. Tech. *18*, 373–386.

Saroj, S.D., Hajare, S., Shashidhar, R., Dhokane, V., Sharma, A., and Bandekar, J.R. (2007). Radiation processing for elimination of *Salmonella typhimurium* from inoculated seeds used for sprout making in India and effect of irradiation on germination of seeds. J. Food Prot. *70*, 1961–1965.

Schlüter, O., Ehlbeck, J., Hertel, C., Habermeyer, M., Roth, A., Engel, K.-H., Holzhauser, T., Knorr, D., and Eisenbrand, G. (2013). Opinion on the use of plasma processes for treatment of foods. Mol. Nutr. Food Res. *57*, 920–927.

Schnabel, U., Niquet, R., Krohmann, U., Polak, M., Schlüter, O., Weltmann, K.-D., and Ehlbeck, J. (2012). Decontamination of microbiologically contaminated seeds by microwave driven discharge processed gas. JASA *1*, 100–106.

Šerá, B., Gajdová, I., Černák, M., Gavril, B., Hnatiuc, E., Kováčik, D., Kříha, V., Sláma, J., Šerý, M., and Špatenka, P. (2012). How various plasma sources may affect seed germination and growth. Proceedings of the International Conference on Optimization of Electrical and Electronic Equipment (OPTIM) 1365–1370.

Šerá, B., Gajdová, I., Šerý, M., and Špatenka, P. (2013). New physicochemical treatment method of poppy seeds for agriculture and food industries. Plasma Sci. Technol. *15*, 935–938.

Selcuk, M., Oksuz, L., and Basaran, P. (2008). Decontamination of grains and legumes infected with *Aspergillus* spp. and *Penicillum* spp. by cold plasma treatment. Biores. Technol. *99*, 5104–5109.

Shiomi, T. (1992). Black rot of cabbage seeds and its disinfection under a hot-air treatment. JARQ *26*, 13–18.

Taormina, P.J., and Beuchat, L.R. (1999). Comparison of chemical treatments to eliminate enterohaemorrhagic *Escherichia coli* O157:H7 on alfalfa seeds. J. Food Prot. *62*, 318–324.

Williams, P.H. (1980). Black rot: A continuing threat to world crucifers. Plant Dis. *64*, 736–742.

Validation of Gas Plasma Sterilization (Importance of ISO documents, ISO TC 198 and 194) 13

Hideharu Shintani

Abstract

ISO 11138-1 and ISO 14161 are the major documents to follow for biological indicator (BI) manufacturers and BI users, respectively. Based on these ISO documents, the spores of *Geobacillus stearothermophilus* ATCC 7953 are used for BI of gas plasma sterilization. In this chapter, the importance of following ISO documents for sterilization validation in the case of gas plasma exposure will be introduced.

ISO documents

I am a Japan delegate of the ISO Technical Committee (TC) 198 and 194. TC 198 covers Sterilization of healthcare products. TC 194 covers Biological and clinical evaluation of medical devices. There are currently no ISO documents covering gas plasma sterilization. I have extensive experience in both the gas plasma sterilization field and as an ISO TC delegate. In addition, I have published ISO-relevant books and papers, so based on these experiences I have prepared this chapter describing the application of ISO requirements to gas plasma sterilization. Gas plasma sterilization is a useful procedure that can easily achieve a sterility assurance level (SAL) of 10^{-6} and material/functional compatibility simultaneously (Chapter 3) (Shintani *et al.*, 2007, 2010; Shintani, 2012; Klaempfl *et al.*, 2012; Venezia *et al.*, 2008). This is because the gas plasma penetration depth is quite shallow (10–20 nm) (Shintani *et al.*, 2007; Shintani, 2012), so material does not deteriorate as easily compared with existing sterilization procedures. Current sterilization validation requires a SAL of 10^{-6} and simultaneous material/functional compatibility. If strict adherence to this requirement were required of the existing sterilization procedures, there would be no sterilization procedures available, so this requirement of simultaneous attainment of SAL of 10^{-6} and material/functional compatibility is in most cases ignored when applied to real sterilization procedures. For example, in our experiments comparing material compatibility among gamma-ray, autoclaving and ethylene oxide gas sterilization, significant deterioration of polyurethane materials was observed (Shintani, 1995). Therefore, if the absolute requirement for sterilization validation were applied (Zheng *et al.*, 2011; Chang *et al.*, 2013; Yuaan *et al.*, 2008; Volny *et al.*, 2007; Kwock *et al.*, 2004; Williams *et al.*, 2004; Olde *et al.*, 2003; Kujipers *et al.*, 2000; Bos *et al.*, 1999; Kawakami *et al.*, 1996; Lin and Cooper, 1995; Courtney *et al.*,

Box 13.1 Four major degradation documents in the ISO 194 document

- Identification and quantification of degradation products from polymeric medical devices is in ISO 10993-13
- Identification and quantification of degradation products from ceramics is in ISO 10993-14
- Identification and quantification of degradation products from metals and alloys is in ISO 10993-15
- Toxicokinetic study design for degradation products and leachables is in ISO 10993-16

1978), no available sterilization procedures would suffice; as a result, sterilization validation is currently a compromise.

In the ISO 194 document, there are four major degradation documents as outlined in Box 13.1. These documents discuss degradation products in the body, not by sterilization, but the mechanism of *in vivo* degradation is quite similar to enzymatic degradation and that of sterilization because bonds with low bonding energy are easily cleaved *in vivo* as well as during sterilization (Table 13.1).

In ISO 198, several sterilization procedures are discussed. Sterilization procedures are individually discussed, but for example chemical indicators (CI) or biological indicators (BI) are common topics of discussion for sterilization validation and routine control. CI documents are presented in ISO 11140-1 to 11140-5 and BI document are in ISO 11138-1 to ISO 11138–5. BI is approved for use in both validation studies and routine control, but CI is not approved for use in validation studies and it is only approved for use in routine control in support of BI. This means that BI and CI differ significantly. Unlike CIs, BIs are absolutely essential to attain sterility assurance as described in several ISO documents.

The biological discussion is documented in ISO 11737-1, 'Sterilization of medical devices-Microbiological methods-Part 1: Determination of a population of microorganisms on products'. CFU of the biological indicator or bioburden in/on the carrier or products can be determined by using ISO 11737–1. Several methods for the recovery of the bioburden from health care products are discussed in ISO 11737-1. ISO 11737-1 discusses only the retrieval of the bioburden from the products, in contrast to ISO 11138-1 and ISO 14161, which deal with BI retrieval from BI carriers such as paper or SUS. Both ISO documents were discussed at TC 198. Tests of sterility performed in the definition, validation and maintenance of a sterilization process is in ISO 11737-2. Aseptic processing, BI use and other issues related to sterility assurance are described in the documents listed in Box 13.2.

I mostly discuss working group 4, which deals with biological indicators (BIs) (see Box 13.3). Among these ISO documents from ISO TC 198 and 194, ISO 11138-1, ISO 14161, ISO 14937, ISO 15882 and ISO 11737-1 are the essential ISO documents to read and comprehend for validation study and routine control.

Table 13.1 Bonding dissociation energy (kcal/mol at 25°C)

	*DH*298 (kcal/mol)	Δf*H*298(R) (kcal/mol)		*DH*298 (kcalmol)	Δf*H*298 (R) (kcal/mol)
Inorganics					
H_2	104.206 ± 0.003	52.103 ± 0.003	$OH^- \rightarrow O^- + H$	110.21 ± 0.07	−33.23 ± 0.07
HF	136.25 ± 0.01	18.83 ± 0.17	$OH^+ \rightarrow O + H^+$	115.2 ± 0.1	59.55 ± 0.02
HCl	103.15 ± 0.03	29.03 ± 0.04	H_2S	91.2 ± 0.1	34.2 ± 0.2
HBr	87.54 ± 0.05	28.62 ± 0.06	SH	84.1 ± 0.2	66.2 ± 0.3
HI	71.32 ± 0.06	26.04 ± 0.08	H-NO	49.5 ± 0.7	21.8 ± 0.1
H-CN	126.3 ± 0.2	105.0 ± 0.7	H-ONO (trans)	79.1 ± 0.2	8.2 ± 0.1
NH_3	107.6 ± 0.1	44.5 ± 0.1	H-ONO_2	101.7 ± 0.4	17.6 ± 0.3
H_2O	118.82 ± 0.07	8.86 ± 0.07	SiH_4	91.7 ± 0.5	47.9 ± 0.6
OH	101.76 ± 0.07	59.55 ± 0.02	GeH_4	83 ± 2	53 ± 2
Hydrocarbons					
CH_4	104.99 ± 0.03	35.05 ± 0.07	CH_2CH-H	110.7 ± 0.6	71.1 ± 0.7
CH_3	110.4 ± 0.2	93.3 ± 0.2	HCC-H	133.32 ± 0.07	135.6 ± 0.2
CH_2	101.3 ± 0.3	142.5 ± 0.2	C_6H_5-H	112.9 ± 0.5	80.5 ± 0.5
CH	80.9 ± 0.2	171.3 ± 0.1	$C_6H_5 \rightarrow o\text{-}C_6H_4 + H$	78 ± 3	106 ± 3
CH_3CH_2-H	101.1 ± 0.4	29.0 ± 0.4	$C_6H_5 \rightarrow m\text{-}C_6H_4 + H$	94 ± 3	122 ± 3
$(CH_3)_2CH$-H	98.6 ± 0.4	21.5 ± 0.4	$C_6H_5 \rightarrow p\text{-}C_6H_4 + H$	109 ± 3	138 ± 3
$CH_3CH_2(CH_3)$ CH-H	98.2 ± 0.5	16.1 ± 0.5	CH_2CHCH_2-H	88.8 ± 0.4	41.4 ± 0.4
$(CH_3)_3C$-H	96.5 ± 0.4	12.3 ± 0.4	$C_6H_5CH_2$-H	89.8 ± 0.6	49.7 ± 0.6
Alcohols					
H-CH_2OH	96.1 ± 0.2	−4.08 ± 0.2	CH_3CH_2O-H	104.7 ± 0.8	−3.6 ± 0.8
CH_3O-H	104.6 ± 0.7	4.3 ± 0.7	$(CH_3)_2CHO$-H	105.7 ± 0.7	−11.5 ± 0.7
CH_3S-H	87.4 ± 0.5	29.8 ± 0.4	$(CH_3)_3CO$-H	106.3 ± 0.7	−20.5 ± 0.7
H-CH_2SH	94 ± 2	36 ± 2	C_6H_5O-H	90 ± 3	−58 ± 3
Peroxides					
HOO-H	87.8 ± 0.5	3.2 ± 0.5	CH_3CH_2OO-H	85 ± 2	−6.8 ± 2.3
CH_3OO-H	88 ± 1	4.8 ± 1.2	$CH_3)_3COO$-H	84 ± 2	−25.2 ± 2.3
Carbonyls					
H-CHO	88.144 ± 0.008	10.1 ± 0.1	H-COOH is ≥	96 ± 1	−46.5 ± 0.7
$CH_3C(O)$-H	89.4 ± 0.3	−2.4 ± 0.3	CH_3COO-H	112 ± 3	−43 ± 3
H-CH_2CHO	94 ± 2	2.5 ± 2.2	C_6H_5COO-H	111 ± 4	−12 ± 4
HCOO-H	112 ± 3	−30 ± 3			

Reproduced from http://www2.chemistry.msu.edu/courses/cem850/handouts/Ellison_BDEs.pdf

Box 13.2 Aseptic processing, BI use and other issues related to sterility assurance

Aseptic processing of health care products

- Part 1: General requirement is in ISO 13408-1
- Part 2: Filtration is in ISO 13408-2
- Part 3: Lyophilization is in ISO 13408-3
- Part 4: Clean-in-place technologies is in ISO 13408-4
- Part 5: Sterilization in place is in ISO 13408-5
- Part 6: Isolator systems is in ISO 13408-6
- Part 7: Alternative processes for medical devices and combination products is in ISO 13408-7.

Sterilization of health care products

Biological indicators

- Guidance for the selection, use and interpretation of results is in ISO 14161
- General requirements for characterization of a sterilizing agent and the development, validation and routine control of a sterilization process for medical devices is in ISO 14937.

Chemical indicators

- Guidance for selection, use and interpretation of results is in ISO 15882.

Biological and chemical indicators

- Test equipment is in ISO 18472.

Box 13.3 Working Group 4: Biological Indicators (BIs)

- Part 1: General requirements is in ISO 11138-1
- Part 2: Biological indicators for ethylene oxide sterilization processes is in ISO 11138-1
- Part 3: Biological indicators for moist heat sterilization processes is in ISO 11138-3
- Part 4: Biological indicators for dry heat sterilization processes is in ISO 11138-4
- Part 5: Biological indicators for low-temperature steam and formaldehyde sterilization process is in ISO 11138-5.

ISO 11138-1 and ISO 14161

For gas plasma sterilization researchers, ISO 11138-1 and ISO 14161 are the essential ISO documents to comprehend. ISO 11138-1 provides general information for BI manufacturing companies and ISO 14161 is specifically for BI users. Therefore, BI manufacturers must obey the ISO 11138 series and BI validation studies and routine control must be carried out following the ISO 11138-1 requirements. According to that document, a BI with an initial population of 10^6 CFU/carrier must be utilized and a sterility assurance level (SAL) of 10^{-6} must be attained, indicating that a 12 log reduction is required for BI manufacturers (ISO 11138-1). In routine control, the use of 10^5 CFU/carrier is approved as an initial population, but the SAL requirement of 10^{-6} is unchanged, so a full 11 log reduction must be attained for routine control according to ISO 11138-1.

On the contrary, ISO 14161 provides the requirements for BI users and describes in detail the half cycle method, overkill method, combined BI/bioburden method and absolute bioburden method. In the half cycle method, the initial population (10^6 CFU/carrier) must be reduced down to the half-cycle window (SAL 5 to SAL 10^{-2}; see Fig. 3.1), indicating a 6 or 8 log reduction, and these figures double. As a whole, a 12 to 16 log reduction is therefore required. No healthcare product manufacturers conduct the half-cycle method due to these very stringent requirements. When using the half-cycle method, it is easy to attain a SAL of 10^{-6}, but material and functional compatibility cannot be attained due to the long exposure time necessary. It is a serious problem if sterilization is successfully completed but the material is no longer useful.

The over-kill method is commonly required for BI manufacturer validation studies, but this method is not always appropriate for BI users (ISO 14161). BI users can conduct other methods besides the over-kill procedure. The BI initial population for the over-kill method is 10^6 CFU/carrier and a SAL of 10^{-6} must be achieved; therefore, a 12 log reduction is required. This is a very difficult requirement for most BI users. Even if a SAL of 10^{-6} can be attained, material and functional compatibility may be impossible to achieve with this method.

According to ISO 11138-1 and ISO 14161, attainment of a SAL of 10^{-6} is absolutely required. However, attainment of material/functional compatibility is required of BI users in ISO 14161, but it is not required for BI manufacturers in ISO 11138-1 because the BI manufacturer has no material being tested. Therefore, the two sets of requirements are clearly different.

The combined BI/bioburden method in ISO 14161 is specifically for BI users. The initial population is the number of *Geobacillus stearothermophilus* ATCC 7953 spores that is approximately equivalent to the bioburden number, so a few CFU/carrier, and a SAL of 10^{-6} is required. In this case, a six log reduction is required. Unfortunately, BIs with this low concentration are not commercially available, so BI users must prepare the BI themselves. At this point they are temporarily considered to be BI manufacturers, and as a result they must follow ISO 11138-1. BI users who make their own BI must be inspected by the authorities as if they were commercial BI manufacturers, but most BI users do not want to undergo these inspections. Commercial BI preparations with a population of more than 10^3 CFU/carrier are available, and in most cases, to avoid inspections, these will be chosen by BI users for use in the combined BI/bioburden method. In this case, reduction from 10^3 CFU/carrier to a SAL of 10^{-6} represents a 9 log reduction. The combined BI/bioburden method is the most popular, and is the most realistic method if the authorities understand the BI/bioburden method requirements. The use of 10^3 CFU/carrier in the same area as 10^6 CFU/carrier means that BI clumping will be less, and therefore no tailing phenomenon is observed. Therefore it is easy to obtain a linear survivor curve and a *D* value is easily defined compared with the overkill or half-cycle methods.

The last option is the absolute bioburden method. The initial population is the actual number of bioburden present (determined based on the average from three products) and the microorganism selected is the most sterilization tolerant among the bioburden microorganisms, so it is not always *Geobacillus stearothermophilus* ATCC 7953 or *Bacillus atrophaeus* ATCC 9372. The possibility of *Bacillus subtilis* may be present as a bioburden (see Table 8.1), but that of *Geobacillus stearothermophilus* is definitely denied because the latter is a definitely thermophilic bacteria growing at 55°C. An initial population of 10^0 CFU/carrier level must be reduced to a SAL of 10^{-6}, so at least a 6 log reduction is required. In this case the BI is also self-made, so the ISO 11138-1 requirements must be met and relevant inspections must be conducted. The absolute bioburden method is the most realistic method, but because the BI is self made, and BI users do not want to undergo the required inspections, in most cases the overkill or combined BI/bioburden method with an initial population of 10^6 CFU/carrier or 10^3 CFU/carrier, respectively, are typically used. The absolute bioburden method is not utilized very often since BI users are reluctant to be inspected by the authorities. The required six log reduction of the absolute bioburden method is the most appropriate and realistic and the use of BI organisms (spores) is not always required. The real exposure time for the absolute bioburden method is significantly less than that for the 9 log or 12 log reduction required by the combined BI/bioburden or overkill methods, respectively. A BI with 10^3 CFU/carrier for the BI/bioburden method has less clumping compared with the 10^6 CFU/carrier BI used in the overkill or half-cycle method, so I recommend the use of the combined BI/bioburden method with 10^3 CFU/carrier commercial BI in place of the 10^6 CFU/carrier BI used for the overkill method (12 log reduction) or half-cycle method (12 log to 16 log reduction). For the commercial BI, I recommend the BI from Merck Co. inoculated on modified SUS carrier because the BI from this company has less clumping as shown in Figs. 1.2 and 1.4.

The use of the overkill method is restricted to BI manufacturers although BI users also can be approved to use it, but BI users need not obey the requirements for the BI manufacturer as described in ISO 14161. When the combined BI/bioburden method is applied, BI clumping is low, so it is easy to attain a SAL of 10^{-6} with a linear survival curve from 10^3 to a SAL of 10^{-6} together with material/functional compatibility. For BI manufacturers and for routine control, the use of 10^5 CFU/carrier of BI in ISO 11138-1 is approved. Such a description of a difference of the initial population between validation studies and routine control is not described in ISO 14161 for BI users. According to the requirements of ISO 14161, validation studies and routine control must use an identical BI for evaluation. According to the current ISO documents, less than a 6 log reduction is not allowed because the approved initial population is 10^0 CFU/carrier and a SAL of 10^{-6} must be obtained; therefore, a at minimum of a 6 log reduction is required, although most engineering researchers are satisfied with 2 log or 3 log reductions, which are not valid.

It is important to mention that chemical indicators (CI) are not approved for use in validation studies. CI manufacturers validate CI characteristics and performance according to the ISO 11140 series and they are also inspected by the authorities. In routine control CI can be utilized as a support for the BI, but use of a CI alone is not approved even for routine control. The BI is the major requirement for sterilization validation and routine control because results using BI and CI do not always coincide.

How to prepare survival curves

Preparation of survival curves is not exactly within the scope of this book, but when considering the cause of tailing, it is an important aspect (Joaquin *et al.*, 2009; Wright *et al.*, 1995). Methods for the preparation of survival curves can be found in ISO 11138-1 Annex B Survival curve (Normative reference). However, in Annex B, there is no description of tailing, but as mentioned above the user must seriously consider the problems caused by clumping when preparing a BI because the coefficient relationship of the prepared survival curve must be more than 0.8 according to ISO 11138-1 Annex B (Normative reference). This requirement indicates no that tailing is allowed from an initial population of 10^6 CFU/carrier to a SAL of 10^{-6}.

In order to prepare a survivor curve, ten-fold dilutions must be repeated to obtain 30–300 CFU/plate (ISO 14161, 11737-1), and the retrieval procedure to obtain the bioburden or BI from the products or carrier material, respectively, must be carried out according to ISO 11737-1. In ISO 11737-1, several sorts of retrieval procedures are presented. The person who conducts the retrieval must validate which method is most appropriate for their products. The retrieval solution that is commonly used is Tween® 80 containing phosphate-buffered saline solution at pH 7.4. This is not defined in ISO 111737-1, but it is the most popular solution when retrieving microorganisms from health care products or BI. The retrieval procedure must be validated individually by referring to ISO 11737-1.

In ISO 11138 Annex C and D (Normative references), the fraction negative methods for *D* value calculation are described; these include the Sperman-Karber procedure and the Stumbo–Murphy–Cochran procedures. The complete fraction negative method is in ISO 14161 Annex C (Normative reference). On the contrary, no survivor curve method is described in ISO 14161; it is only described in ISO 11138-1 B (Normative reference).

How to calculate *D* values

There are two methods to calculate *D* values. One is by using the survivor curve method, which is in Annex B (Normative reference) of ISO 11138-1 and the others are fraction negative methods in ISO 14161 Annex C (normative reference). In regard to the survivor curve method in ISO 11138-1 Annex B, it is defined that the reduction from an initial population of 10^6 CFU to a SAL of 10^{-6} must be linear with a coefficient correlation greater than 0.8, which is quite a stringent requirement.

Fraction negative methods are the Sperman–Karber procedure and Stumbo–Murphy–Cochran procedures. They are described in Annex C (Normative reference) of ISO 14161 and are for BI users. A description of methods for the calculation of *D* values is outside of scope of this book, but BI users need to keep in mind that less than 10^6 CFU can be used as an initial population because BI users must obey ISO 14161, not ISO 11138-1.

Conclusion

Many ISO documents must be read and understood in order to conduct sterilization validation and routine control. If published papers and books do not follow ISO requirements, these books and papers are considered invalid. Most of the engineering researchers' papers and books are invalid because their survivor curves do not attain a SAL of 10^{-6}, primarily because of the use of BI with clumping. The SALs presented by the engineering researchers' are 10^2 to 10^3 at most, which differs significantly from 10^{-6}, which is definitely required by the ISO documents. Therefore, the data in their books and papers are invalid and unreliable.

In order to avoid the tailing phenomenon from clumping, less than 10^6 CFU/carrier BI (such as 10^3 CFU/carrier BI) can be used to attain a SAL of 10^{-6}. It should be relatively easy to attain a SAL of 10^{-6} using the combined BI/bioburden method in ISO 14161 and BI users should obey this ISO document, not ISO 11138-1, which is only for BI manufacturers. If BI users, including researchers do not follow the requirements of ISO 14161, their papers and books are useless and invalid.

Acknowledgement

This chapter is reproduced from a previous article (Hideharu Shintani, 2015, Pharmaceut Reg Affairs 4:e150. doi: 10.4172/2167-7689.1000e150) written by the same corresponding author (Hideharu Shintani) and distributed under the terms of the Creative Commons Attribution License (CCAL). It should be noted that CCAL permits unrestricted use, distribution and reproduction in any medium.

References

Bos, G.W., Scharenborg, N.M., Poot, A.A., Engbers, G.H., Beugeling, T., van Aken, W.G., and Feijen, J. (1999). Blood compatibility of surfacers with immobilized albumin–heparin conjugated and effect of endothelial cell seeding on platelet adhesion. J. Biomed. Mater. Res. *47*, 279–291.

Chang, Y.Y., Huang, H.L., Lai, C.H., Hsu, J.T., Shieh, T.M., Wu, A.Y., and Chen, C.L. (2013). Analyses of antibacterial ctivity and cell compatibility of titanium coated with Zr-C-N fil., PLoS One. 2013;8(2):e56771. Doi;1371

Courtney, J.M., Park, G.B., Prentice, C.R., Winchester, J.F., and Forbes, C.D. (1978). Polymer modification and blood compatibility. J. Bioeng. *2*, 241–249.

Joaquin, J.C., Kwan, C., Abramzon, N., Vandervoort, K., and Brerlles-Marino, G. (2009). Is gas discharge plasma a bew solution to the old problem of biofilm inactivation? Microbiol.*155*, 724–732.

Kawakami, H., Nagaoka, S., and Kubota, S. (1996). Gas transfer and in vitro and in vivo blood compatibility of a fluorinated polyimide membrane with an ultrathin skin layer. ASAIO J. *42*, M871–876.

Klaempfl, T.G., Isbary, G., Shimizu, T., Li, Y.-F., Zimmermann, J.L., Stolz, W., Schiegel, J., Morfill, G.E., and Schmidt, H.-U. (2012). Cold atmospheric air plasmas sterilization against spores and other microorganisms of clinical interest. Appl. Environ. Microbiol. *78*, 5077–5082.

Kujipers, A.J., van Wachem, P.B., van Luyn, M.J., Plantinga, J.A., Engbers, G.H., Krijgsveid, J., Zaak, S.A., Dankert, J., and Feijen, J. (2000). In vivo compatibility and degradation of crosslinked gelatin gels incorporated in knitted Dacron. J. Niomed. Master. Res. *51*, 136–145.

Kwock, S.C., Yang, P., Wang, J., Liu, X., and Chu, P.K. (2004). Hemocompatibility of nityrogen-doped, hydrogen-gree diamond-like carbon prepared by nitrogen plasma immersion ion implantation-deposition. J. Biomed. Mater. Res. A. *70*, 107–114.

Lin, J.C., and Cooper, S.L. (1995). Surface characterization and ex vivo blood compatibility study of plasma-modified small diameter tubing; effect of sulphur dioxide and hexamethyldisiloxane plasmas. Biomaterials *16*, 1017–1013.

Olde Riekerink, M.B., Claase, M.B., Engbers, G.H., Grijpma, D.W., and Feijen, J. (2003). Gas plasma etching of PEO/PBT segmented block copolymer films. J. Biomed. Mater. Res. *65*, 417–428.

Shintani, H., Sakudo, A., Burke, P., and McDonnell, G. (2010). Gas plasma sterilization of microorganisms and mechanisms of action, Exp. Ther. Med. *1*, 731–743.

Shintani, H., Shimizu, N., Imanishi, Y., Sekiya, T., Tamazawa, K., Taniguchi, A., and Kido, A. (2007). Inactivation of microorganisms and endotoxins by low ytemperature nitrogen gas plasma exposure. Biocontrol. Sci. *12*, 131–143.

Shintani, H. (2012). Inactivation of prion and endotoxins by nitrogen gas plasma exposure. Pharmaceut. Anal. Acta 3, http://dx.doi.org/10.4172/2153–2435.1000177

Shintani, H. (1995). The relative safety of gamma-ray, autoclave and ethylene oxide gas sterilization of thermosetting polyurethane. Biomed. Instrum. Technol. *29*, 513–519.

Venezia, R., Orrico, M., Houston, E., Yin, S-M., and Naumova, Y.Y. (2008). Lethal activity of nonthermal plasma sterilization against microorganisms. http://www.jstor.org/stable/10.1086/588003.
Volny, M., Elam, W.T., Ratner, B.D., and Turecek, F. (2007). Enhanced in –vitro blood compatibility of 316L stainless steel surfaces by reactive landing of hyaluronan ions. J. Biomed. Mater. Rers. B. Appl. Biomater. *80*, 505–510.
Williams, R.L., Wilson, D.J., and Rhodes, N.P. (2004). Stability of plasma-treated silicone rubber and its influence on the interfacial aspects of blood cpmpatibility. Biomaterials, *25*, 4659–4673.
Wright, A.M., Hoxey, E.V., Soper, C.J., and Davies, D.J. (1995). Biological indicators for low temperature steam and formaldehyde sterilization: the effect of defined media on sporulation, growth index and formaldehyde resistance of spores of Bacillus stearothermophilus strains. J. Apopl. Bacteriol. *79*, 235–238.
Yuan, Y., Liu, C., and Yin, M. (2008). Plasma polymerized *n*-butyl methacrylate coating with potential for re-endothelialization of intravascular stent devices. J. Mater. Sci. Materr. Med. *19*, 2187–2196.
Zheng, Q., Yue, P.F., Wu, B., Hu, P.Y., Wu, Z.F., and Yang, M. (2011). Pharmacokinetics comparative study of a novel Chinese traditional herbal formula and its compatibility. J. Ethnopharamacol. *137*, 221–225.

Misinterpretation of Microbiological Data on Gas Plasma Sterilization: Avoiding the Pitfalls

14

Hideharu Shintani

Abstract

In this chapter, we discuss avoidance of clumping, tailing phenomena, and methods for attaining a SAL of 10^{-6} using a biological indicator (BI). As tailing of a survivor curve is often caused by clumping of the BI, appropriate techniques and confidential know-how to avoid such clumping is required. When such a tailing phenomenon is observed, a SAL of 10^{-6} cannot be attained, and therefore no *D* value (decimal reduction value) can be determined and the exposure time for a 9 or 12 log reduction remains undefined. The *D* value must be determined from the straight line of a 9 log or 12 log reduction survivor curve, and there can only be one *D* value per microorganism; there is never more than one *D* value per one microorganism.

Introduction

Several papers and books on gas plasma sterilization have been published, mostly by engineering researchers. Due to their insufficient knowledge of sterilization and microbiology, their papers and books contain many misinterpretations of data in the figures and tables and descriptions in the text (e.g. Rossi and Kylian, 2012). A typical example is the curved survival curve (see Fig. 1.3A). In this chapter, I would like to convey the correct information on sterilization, and how to conduct sterilization validation studies and routine control for BI users and BI manufacturers.

The requirements for sterilization validation studies and routine control for BI manufacturers and BI users are different. BI manufacturers should use 10^6 CFU/carrier BI and must attain a SAL of 10^{-6} in validation studies (ISO 11138-1). In routine control, it is approved to use 10^5 CFU as an initial population, but a SAL of 10^{-6} must be attained according to ISO 11138-1. For BI users, the use of various initial populations is approved according to ISO 14161, although the initial population used in validation studies and routine control should be the same. This information is described in detail in Chapter 13.

Sterilization

Sterilization kills all types of bacterial spores and vegetative cells, which means the material is totally free of bioburden. The bioburden is the number and types of viable microorganisms in/on the products, or the so-called 'viable contaminants'. In contrast, disinfection only kills vegetative cells, but not bacterial spores. Decontamination is the removal of spores and vegetative cells by a process that does not kill them. The definitions of sterilization, disinfection and decontamination are described in the book edited by Sakudo and Shintani (2011).

Sterilization is defined by the ISO 11138 series and ISO 14161. However, no ISO documents on gas plasma sterilization are currently available. Recent Pub Med searches with the key words ISO and gas plasma sterilization, did not result in any matches. As described in Chapter 13, the ISO 11138 series is addressed to BI manufacturers and ISO 14161 is for BI users. According to ISO 11138-1, a BI with an initial population of 10^6 CFU/carrier must be used and a sterility assurance level (SAL) of 10^{-6} must be attained, indicating that a 12 log reduction is required in validation studies. This requirement is only for BI manufacturers, not for BI users, so BI users need not follow the 12 log reduction procedure. For example, 10^3 CFU/carrier BI has been approved for use by BI users (Combined BI/bioburden method) and a SAL of 10^{-6} is required, which is a 9 log reduction. Because of the lower population for a BI with 10^3 CFU/carrier, clumping is less of a problem compared with a 10^6 CFU/carrier BI; this can be experimentally confirmed by using scanning electron microscopy (SEM). If there is less clumping, the survivor curve is more likely to be a straight line (see Figs. 1.3B and 1.4) and any tailing phenomenon will be avoided (Ahn *et al.*, 2007; Geeraerdd *et al.*, 2006; Moison *et al.*, 2001, 2002; Rossi and Kylian, 2012). If the survivor curve shows tailing, a SAL of 10^{-6} cannot be attained, the *D* (decimal reduction) value, time or dose to decrease one log reduction, cannot be determined, and the sterilization exposure time to achieve a 9 log or 12 log reduction cannot be obtained without any information regarding a *D* value. In other words, no valid results can be attained when there is clumping of the BI and tailing of the survivor curve. The *D* value must be determined from the straight slope of a 12 log reduction survivor curve, and there is only one *D* value per one microorganism. The inactivation kinetics follows a first order equation. Even if tailing occurs, there cannot be more than one *D* value per microorganism and inactivation kinetics cannot be second or third order when restricted to BIs such as *Geobacillus stearothermophilus* ATCC 7953 or *Bacillus atrophaeus* ATCC 9372 because they are spore forming bacteria with straight survivor curves against all sorts of sterilization processes tested.

Most engineering researcher's data show curved survivor curves and report that inactivation kinetics is first or second order, indicating more than two *D* values per each microorganism. Their data are invalid according to the requirements of ISO 14161, because the *D* value must be determined from the straight slope from the initial population of 10^6 CFU/carrier to a SAL of 10^{-6}, indicating a 12 log reduction survivor curve. This indicates that the inactivation kinetics follows a first order equation because the use of *Geobacillus stearothermophilus* ATCC 7953 spores as the BI should not result in a tailing phenomenon. In sterilization validation, a 6 log reduction is not approved except in the absolute bioburden method in ISO 14161 for BI users.

D values can be obtained by the fraction-negative method using the Stumbo–Murphy–Cochran procedure and Spermann–Kerber procedure (ISO 14161). These methods require that from an initial population of 10^6 CFU/carrier to a SAL of $5–10^{-2}$, the survivor curve (see Fig. 3.1) must be a straight line, and clumping should not be present. As a description of the methods for calculation of *D* values is outside the scope of this chapter, please refer to ISO 14161 and ISO 11138-1 and Chapter 13. Procedures for the calculation of *D* values are more clearly described in ISO 14161.

A BI for gas plasma sterilization is not currently defined in any ISO TC 198 documents. The BI is defined the most sterilization-tolerant non-pathogenic microorganism, and is generally a bacterial spore former. From the available experimental data, *Geobacillus stearothermophilus* ATCC 7953 is the most appropriate candidate. In some cases *B. atrophaeus* ATCC 9372 has been used, but its bacterial spore are less tolerant than that of *Geobacillus stearothermophilus* ATCC 7953 (Deng, *et al* 2006; Shintani *et al.*, 2007). Most engineering researchers typically use bacteria that do not form endospores, such as *Escherichia coli*, *Legionella* spp., and so on, but these microorganisms are very susceptible to sterilization compared with bacterial spores, so their use is invalid. Even if sterilization studies confirm that *E. coli* or *Legionella* spp. can be disinfected, what about *Bacillus cereus*, which is a pathogenic spore forming microorganism? This is a major concern because there is a reasonable chance that bacterial spore formers may be present as a bioburden (as an example, see Table 8.1). How can such studies address whether contaminants such as *B. cereus* can be sterilized or not? In order to sterilize pathogenic bacterial spores it is necessary to use an appropriate BI for confirmation, and the BI must be the most tolerant bacterial spore to the sterilization procedure because pathogenic bacterial spores present as a bioburden can also be speculated to be killed. Therefore, sterilization, not disinfection or decontamination must be carried out. If sterilization is attained using an appropriate BI, it is reasonable to consider that other pathogenic microorganisms of vegetative cell type or even bacterial spores would also be killed by the same sterilization process, and further experimentation is unnecessary. This means that for gas plasma sterilization the use of *Geobacillus stearothermophilus* ATCC 7953 as the BI is required and use of *Bacillus atrophaeus* ATCC 9372 is inappropriate.

Gas plasma sterilization has a quite shallow penetration depth of approximately 10–20 nm (Shintani *et al.*, 2007). Because penetration is so shallow, only one layer of bioburden can be sterilized, and the healthcare products being sterilized remain undamaged. In validation studies, a SAL of 10^{-6} and material/functional compatibility must be simultaneously attained. Among the existing sterilization procedures, no sterilization methods can currently achieve simultaneous achievement of a SAL of 10^{-6} and material/functional compatibility. If this requirement is strictly enforced for the existing sterilization procedures, there are no compliant sterilization procedures available, and therefore the use of the existing sterilization procedures is the result of a compromise. As gas plasma sterilization can successfully comply with both sterilization requirements, the current compromise will be problematic, as other sterilization methods will then be strictly required to simultaneously attain both a SAL of 10^{-6} and material/functional compatibility, which is an impossible requirement for existing sterilization procedures.

Engineering researchers have insufficient knowledge of sterilization and microbiology; therefore, these should cooperate with microbiologists and chemists to correctly evaluate their experimental results. Currently, many engineering researchers misinterpret the meaning of a six log reduction and ignore the need to attain a SAL of 10^{-6}, as seen by the tailing phenomenon in their survival curves due to clumping (Shintani *et al.*, 2010; Shintani and McDonnell, 2011). These researchers need to read ISO 14161 and comprehend the usefulness and importance of using an appropriate BI to conduct sterilization validation.

Engineering researchers' understanding of 6 log reduction is from an initial population of 10^6 CFU/carrier to SAL of 10^0, but this is wrong. The correct requirement is the reduction of an initial population of 10^0 CFU/carrier to a SAL of 10^{-6}. As mentioned in Chapter 13, a SAL of 10^0 CFU is the bioburden level. For this purpose a linear survivor curve is required at least from a SAL of 10^0 to an initial population of 10^6 CFU/carrier (see Figs. 1.4 and 3.1). If a tailed survivor curve is obtained, the experiment is not valid, sterilization has not been attained, and no useful information such as the *D* value can be determined. A curved line at around a SAL of 10^3, which is commonly seen in engineering researchers' papers, provides no useful information. They obtain one *D* value from the initial straight line and another *D* value from the next curved line, which is incorrect. Only one *D* value exists per one microorganism.

Conclusion

Obtaining a SAL of 10^{-6} and material and functional compatibility is a difficult task for existing sterilization procedures. However, in gas plasma sterilization, this requirement can be easily attained because of the shallow penetration depth of the sterilization factors, which include radicals and metastables; therefore, the BI must be completely free from clumping to avoid any tailing phenomenon due to clumping.

If the survivor curve shows tailing, a SAL of 10^{-6} cannot be reached and therefore the *D* value cannot be defined and the exposure time to attain a 9 or 12 log reduction cannot be determined. Under these conditions sterilization cannot be attained. As engineering researchers have an insufficient understanding of sterilization and microbiology, they should cooperate with microbiologists and chemists to evaluate and interpret their experimental results.

Acknowledgement

This chapter is reproduced from a previous article (Hideharu Shintani, Pharmaceutical Reg Affairs, 2015,4:138. doi: 10.4172/2167-7689.1000138) written by the same corresponding author (Hideharu Shintani) and distributed under the terms of the Creative Commons Attribution License (CCAL). It should be noted that CCAL permits unrestricted use, distribution and reproduction in any medium.

References

Ababouch, L.H., Grimit, L., Eddafry, R., and Busta, F.F. (1995). Thermal inactivation linetics of Bacillus subtilis spores suspended in buffer and in oils. J. Appl. Bacteriol. *78*, 669–676.

Ababouch, L., Dikrta, A., and Busta, F.F. (1987). Tailing of survivor curves of clostridial spores heated in edible oils. J. Appl. Bacteriol. *62*, 503–511.

Ahn, J., Balasubrameniam, V.M., and Yousef, A.E. (2007). Inactivation kinetics of selected aerobic and anaerobic bacterial spores by pressure assisted thermal processing. Int. J. Food Microbiol. *113*, 321–325.

Deng, X., Shi, L., and Kong, M.G. (2006). Physical mechanisms of inactivation of Bacillus subtilis spores using cold atmospheric plasmas. IEEE Trans. Plasma Science *34*, 1310–1316.

Furukawa, S., Narisawa, N., Watanabe, T., Kawarai, T., Myozen, K., Okazaki, S., Ogihara, H., and Yamasaki, M. (2005). Formation of the spore clumps during heat treatmenty increase the heat resistance of bacterial spores. Int. J. Food Microbiol. *102*, 107–111.

Geeraerd, A.H., valdramidis, V.P., and Van Impe, J.F. (2006). GInaFIT. A freeware tool to assess non-;log-linear microbial survival curve. Int. J. Food Microbiol. *102*, 95–105.

McDonnell, G.E. (2013). Gas Plasma Sterilization in Principles and Practice of Disinfection, Preservation and Sterilization. Fraise, A.P., Maillard, J.-Y., and Sattar, S.A., eds (Wiley-Blackwell), 5th edn, Oxford, UK, pp.333–342.

Moison, M., Barbeau, J., Crevier, M.-C., Pelletier, J., Philip, N., and Saoudi, N. (2002). Plasma sterilization. Methods and mechanisms, Pure Appl. Chem., *74*, 349–358.

Moisan, M., Barbeau, J., Moreau, S., Pelletier, J., Tabrizian, M., and Yahia, L.H. (2001). Low temperature sterilization using gas plasma: a review of the experiments and an analysis of the inactivation mechanisms. Int. J. Pharm. *226*, 1–21,

Morent, R., and De Geyter, N. (2011). Inactivation of Bacteria by Non-Thermal Plasmas in Biomedical Engineering-Frontiers and Challenges, Reza Fazel Renzai ed., CC BY NC-SA 3.0 license.

Rossi, F., and Kylian, O. (2012). Sterilization and decontamination of surfaces by plasma discharges in Sterilization of biomaterials and medical devices, Lerouge, S., and Simmons, A., eds (WoodHead Publishing, Cambridge), UK, pp. 117–150.

Shintani, H., and McDonnell, G. (2011). Several Points to Consider When conducting Plasma Experiments in Sterilization and Disinfection by Plasma, Sterilization Mechanisms, Biological and Medical Applications, Sakudo, A., and Shintani, H., eds (NOVA Science Publishers Ltd). pp. 223–228, New York.

Shintani, H., and McDonnell, G. (2011). Definition of Sterilization, Disinfection, Decontamination and Antisepsis. Sakudo, A., and Shintani, H., eds (NOVA Science Publishers Ltd). pp. 7–18, New York.

Shintani, H., Sakudo, A., Burke, P., and McDonnell, G. (2010). Gas plasma sterilization of microorganisms and mechanisms of action, Exp. Ther. Medicine (2010). *1*, 731–738.

Shintani, H., Shimizu, N., Imanishi, Y., Sekiya, T., Tamazawa, K., Taniguchi, A., and Kido, N. (2007). Inactivation of microorganisms and endotoxins by low temperature nitrogen gas plasma exposure, Biocontrol Science *12*, 131–143.

Tessarolo, F., Caola, I., Nollo, G., Antolini, R., Guarrera, G.M., and Caciagli, P. (2006). Efficiency in endotoxin removal by a reprocessing protocol for electrophysiology catheters based on hydrogen peroxide plasma sterilization. Int. J. Hyg. Environ. Health *209*, 557–565.

Future Perspectives and Trends in Gas Plasma Sterilization

15

Hideharu Shintani

Abstract

Gas plasma sterilization has a shallow penetration depth of approximately 10–20 nm. Thus, material and functional compatibility can be easily attained. Sterility assurance of a 12 log reduction is not always required of BI users (ISO 14161). Indeed, a 12 log reduction is quite difficult to achieve with the current gas plasma process, because a tailing phenomenon in the survivor curve can occur due to clumping of the biological indicator. Similarly, a 6 log reduction from 10^6 CFU/carrier to a SAL of 100 is also difficult to attain as is often seen in the engineering research. Nonetheless, a 12 log reduction is required in sterilization validation studies for BI manufacturers according to ISO 11138-1. However, simultaneous attainment of material/functional compatibility is not required of BI manufacturers in ISO 11138-1 because there are no materials being tested. When using the overkill method, a 12 log reduction is required together with material/functional compatibility for BI users according to ISO 14161. To achieve this requirement, deeper penetration of gas plasma sterilization will be needed in the future, but the penetration depth must be limited in order to maintain material/functional compatibility.

Future trends

There are several other sterilization procedures beside gas plasma sterilization. The existing sterilization procedures have deeper penetration capabilities, so even if the biological indicator (BI) has any clumping, no tailing phenomenon in the survivor curve is observed. But material/functional compatibility is quite difficult to attain for these existing sterilization procedures due to the deeper penetration depth. For example, ethylene oxide gas and gamma-ray irradiation sterilization penetrate more than 3 and 10 m, respectively, and therefore clumping of the BI or multiple layers of bioburden are not problematic for these sterilization procedures. On the contrary, the penetration depth of gas plasma sterilization procedure is quite shallow, at around 10–20 nm (Shintani, 2007); therefore, it is sufficient to kill one layer of bioburden without reduction of material/functional compatibility but insufficient to kill multilayers of bioburden or clumped BI (Shintani *et al.*, 2007, 2010, 2011; Abreu *et al.*, 2013; Stoffels *et al.*, 2004; Baier *et al.*, 1992; Vandervoort and Brelles-Marino, 2014; Idllbi *et al.*, 2013; Alkawareek *et al.*, 2012; Tessarolo *et al.*, 2006; Traba *et al.*, 2013; Fricke *et al.*, 2012; Brelles-Marino, 2012; Cotter *et al.*, 2011; Joaquin, 2009). In addition, for example fungi, which are large and often in more than one layer, are present as the bioburden they may be difficult to kill with the 10–20 nm penetration capacity of gas plasma exposure. This means that a somewhat deeper penetration would be useful to attain both a SAL of 10^{-6} and material/functional compatibility. A twelve log reduction is required in sterilization validation together with material/functional compatibility when using the overkill method (ISO 14161). A twelve log reduction using gas plasma sterilization has not been reported in any published studies to date. More than a 6 log reduction with a linear survivor curve were reported only by Shintani, as presented in Fig. 1.4. This is due to the use of clump-free BI. To attain a 12 log reduction with ease by gas plasma sterilization, a penetration of at least 1000 nm (1 μm) would be required. Even if the penetration depth were 1 μm, material and functional compatibility would be maintained and sterility assurance would be increased to a 12 log reduction rather than a 6 log or 9 log reduction. A 6 log reduction is for the absolute bioburden method and a 9 log reduction is required for the combined BI/bioburden method (ISO 14161; see Chapter 13). These requirements are only for BI users and are only discussed in ISO 14161, but not in ISO 11138-1 as described in Chapter 13. With the exception of gas plasma exposure, the currently available sterilization procedures have penetration depths of a few metres, so even if the BI has some clumping, no tailing phenomenon is observed in the survivor curves for the existing sterilization procedures. But significant failure of material/functional compatibility attainment often occurs when using the existing sterilization procedures due to the deeper penetration depth (Shintani, 1995).

Gas plasma sterilization will fail to attain sterility assurance when the sterilization target consists of highly contaminated items with a large bioburden such as dental materials and surgical devices; in these cases a 10–20 nm penetration depth is too shallow. Successful sterilization of such items would require an increase in penetration depth of at least 100-fold. Even if future gas plasma sterilization innovations resulted in a 1–10 μm penetration depth, material/functional compatibility would be maintained together with the achievement of a SAL of 10^{-6} as required by ISO 14161, the authorities and GMP (good manufacturing practice, http://www.ispe.org/gmp-resources).

The sterilization factors present in gas plasma sterilization have been proposed to be metastables or photons. By accelerating the metastables or photons before application to the target items, the penetration depth may become deeper. In actuality, a 10–20 nm penetration depth has been demonstrated to be sufficient for BI sterilization as long as the BI is free from clumping. Under real circumstances, however, several unexpected situations must be kept in mind to avoid failure of gas plasma sterilization. For example, the bioburden may have multiple layers or may contain large cells like fungi, or in some cases contain complex materials like biofilms (Vandervoort and Brelles-Marino, 2014; Trabe *et al.*, 2013; Idlibi *et al.*, 2013; Joaquin *et al.*, 2009: Niemira *et al.*, 2014; Sun *et al.*, 2013; Cotter *et al.*, 2011; Matthes *et al.*, 2014; Chang *et al.*, 2013; Ojano-Dirain and Antonelli, 2011; Ermolaeva *et al.*, 2011, 2012; Zelaya *et al.*, 2010; Brelles-Marino, 2010) or waterborne microorganisms including viable but non-culturable microorganisms (VBNC; Hayes *et al.*, 2013; Brelles-Marino, 2012; Rowan *et al.*, 2007). Sterilization of the BI is a model sterilization situation, so complex and highly contaminated targets must be carefully considered in order to avoid sterilization failure in real-life sterilization situations. A 12 log reduction and material/functional compatibility is required for real sterilization situations according to ISO 14161 and 14937, and for that purpose deeper penetration will be required for future improvement of gas plasma sterilization.

Gas plasma exposure is not suitable for liquids, oils, powders, or biological tissues such as highly contaminated items. This is because gas plasma sterilization has inferior penetrating ability as mentioned above; therefore, it is applicable for sterilization of the bioburden on human skin and has no hazardous effects on humans. The advantages of gas plasma sterilization are that it is fast, it is carried out at low-temperature and it has lower penetration with no toxic gas residue generated.

Furthermore, although it has been predicted that the factors contributing to the gas plasma sterilization process may be metastables or photons, this is only speculation, and therefore the real factors must be defined quantitatively in future studies.

Conclusion

The depth of gas plasma sterilization penetration is quite shallow at around 10–20 nm, so it is easy to attain material/functional compatibility, and a 12 log reduction under standard conditions for the BI. However, in real world sterilization situations, a 12 log reduction is very difficult to attain simultaneously with material/functional compatibility due to the existence of admixtures and the presence of highly contaminated items. If the penetration depth were increased up to 1 to 10 μm, the requirements for sterilization validation in ISO 14161 could be attained with ease, indicating that material/functional compatibility and a SAL of 10^{-6} can be attained with only an increase in penetration depth of 100-fold. Therefore, the development of gas plasma sterilization processes with deeper penetration depths may be essential in the future. Currently, gas plasma sterilization is restricted to simple matrix materials, and achievement of deeper penetration is desired so that complex matrix material can also be sterilized. In addition, if the penetration depth were increased, sterilization chambers could be made larger, so that more items could be sterilized simultaneously.

References

Abreu, A.C., Tavares, R.R., Borges, A., Merqulhao, F., and Simoes, M. (2013). Current and emergent strategies for disinfection of hospital environments. J. Antimicrob. Chemother. *66*, 2718–2732.

Alkawareek, M.Y., Algwari, Q.T., Gorman, S.P., Graham, W.G., O'Connell, D., and Gilmore, B.F. (2012). Application of atmospheric pressure nonthermal plasma for the in vitro eradication of bacterial biofilms. FEBS Immunol. Med. Microbiol. *65*, 381–384.

Baier, R.E., Carter, J.M., Sorensen, S.E., Meyer, A.E., McGowan, B.D., and Kasprzak, S.A. (1992). Radiofrequency gas plasma (glow discharge) disinfection of dental operative instruments, including handpieces. J. Oral. Implantol. *18*, 236–242.

Brelles-Marino, G. (2010). Bacterial biofilm inactivation by gas-discharge plasmas. In Biological and Environmental Applications of Gas Discharge Plasmas. Brelles-Marino, G. ed. (NOVA Science Publishers), New York. pp. 75–98.

Brelles-Marino, G. (2012). Induction of a viable-but-non-culturable state in bacteria treated with gas discharge plasma. J. Appl. Microbiol. *112*, 412–413.

Chang, Y.Y., Huang, H.L., Lai, C.H., Hsu, J.T., Shieh, T.M., Wu, A.Y., and Chen, C.L. (2013). Analysis of antibacterial activity and cell compatibility of titanium coated with a Zr-C-N film. PLoS One *8*, e56771. Doi:10.1371.

Cotter, J.J., Maguire, P., Soberon, F., Daniels, S., O'Gara, J.P., and Cassey, E. (2011). Disinfection of meticcillin-resistant Staphylococcus aureus and *Staphylococcus epidermidis* biofilms using a remote non-thermal gas plasma. J. Hosp. Infect. *78*, 204–207.

Ermolaeva, S.A., Varfolomeev, A.F., Chernukha, M.Y., Yurov, D.S., Vasllev, M.M., Kaminskaya, A.A., Moisenovich, M.M., Romanova, J.M., Murashev, A.N., Selezneva, I.I., Shimizu, T., Sysolyatina, E.V., Shaginyan, I.A., Petrov, O.F., Mayevsky, E.I., Fortov, V.E., Morfill, G.E., Naroditsky, B.S., and Fricke, K., Koban, I., Tresp, H., Jablonowski, L., Schroeder, K., Kramer, A., Weltmann, K.D., Woedtke, T., and Kocher, T. (2012). Atmospheric pressure plasma; a high-performance tool for the efficient removal of biofilms. PLoS One 7(8):e42539 doi;10.1371/journal pone 0042539.

Ermolaeva, S.A., Varfolomeev, A.F., Chernukha, M.Y., Yurov, D.S., Vasllev, M.M., Kaminskaya, A.A., Moisenovich, M.M., Romanova, J.M., Murashev, A.N., Selezneva, I.I., Shimizu, T., Sysolyatina, E.V., Shaginyan, I.A., Petrov, O.F., Mayevsky, E.I., Fortov, V.E., Morfill, G.E., Naroditsky, B.S., and Gintsburg, A.L. (2011). Bacteriocidal effects of non-thermal argon plasma in vitro, in biofilm and in the animal model of infected wounds. J. Med. Microbiol. *60*, 75–83.

Fricke, K., Koban, I., Tresp, H., Jabionowski, L., Schroeder, K., Kramer, A., Weltmann, K.D., von Woedtke, T., and Kocher, T. (2012). Atmospheric pressure plasma: a high-performance tool for the efficient removal of biofilms. PLoS One. 2012:7(8):e42539:doi:10.1371

Gintsburg, A.L. (2011). Bacteriocidal effects of non-thermal argon plasma in vitro, in biofilm and in the animal model of infected wounds. J. Med. Microbiol. *60*, 75–83.

Hayes, J., Kirf, D., Garvey, M., and Rowan, N. (2013). Disinfection and toxicological assessments of pulsed UV and pulsed-plasma gas-discharge treated-water containing the waterborne protozoan enteroparasite *Cryptosporidium parvum*. J. Microbiol. Methods *91*, 325–337.

Idlibi, A.N., Al-Marrewi, F., Hannig, M., Lehmann, A., Rueppell, A., Schindler, A., Jentsch, H., and Rupf, S. (2013). Destruction of oral biofilms formed in situ on machined titanium (Ti) surfaces by cold atmospheric plasma. Biofouling *29*, 369–379.

Joaquin, J.C., Kwan, C., Abramzon, N., Vandertvoort, K., and Breelles-Marifio, G. (2009). Is gas-discharge plasma a new solution to the old problem of biofilm inactivation? Microbiology, *155*, 724–732.

Matthes, R., Huebner, N.O., bender, C., Koban, I., Horn, S., Bekeschus, S., Weltmann, K.D., Kocher, T., Kramer, A., and ssadian, O. (2014). Efficacy of different carrier gas for barrier discharge plasma generation compared to chlorohexidine on the survival of *Pseudomonas aeruginosa* embedded in biofilm in vitro. Skin. Pharmacol. Physiol. *27*, 148–157.

Niemira, B.A., Boud, G., and Sites, J. (2014). Cold plasma rapid decontamination of food contact surfaces contaminated with Salmonella biofilms. J. Food Sci. *79*, M917–922.

Ojano-Dirain, C., and Antonelli, P.J. (2011). Pseudomonas biofilm formation after *Haemophilus* infection. Ototaryngol. Head Neck Surg. *145*, 470–475.

Shintani, H., and McDonnell G.E. (2011). Several Points to Consider When conducting Plasma Experiments In Sterilization and Disinfection by Plasma, Sterilization Mechanisms, Biological and Medical Applications, A. Sakudo and H. Shintani, eds (NOVA Science Publishers Ltd.), New York.

Shintani, H., Sakudo, A., Burke, P., and McDonnell, G.E. (2010). Gas plasma sterilization of microorganisms and mechanisms of action, Exp. Ther. Medicine (2010) *1*, 731–738.

Shintani, H., Shimizu, N., Imanishi, Y., Sekiya, T., Tamazawa, K., Taniguchi, A., and Kido, N. (2007). Inactivation of microorganisms and endotoxins by low temperature nitrogen gas plasma exposure, Biocontrol Sci., *12*, 131–143.

Shintani, H. (1995). Ther relative safety of gamma-ray, autoclaver and ethylene oxide gas sterilization of thermosetting polyurethane. Biomed. Instrum. Terchnol. *29*, 513–519.

Shintani, H. (1995). Formation and elution of toxic compounds from sterilized medical products: methylenedianiline formation in polyurethaner. J. Biomater. Appl. *10*, 23–58.

Stoffels, E., Kieft, I.E., Sladek, R.E., Laan, E.P., and Slaaf, D.W. (2004). Gas plasma treatment: a new approach to surgery? Crit. Rev. Biomed. Eng. *32*, 427–460.

Sun, K., Yang, X., Ye, G., Pan, H., and Wang, J. (2013). Evaluation of two different cold plasma treatments on root canal infected with enterococcus faecalis biofilm [in Chinese]. Hua Xi Kou Qiang Yi Xue Zhi. *31*, 195–198.

Tessarolo, F., Caola, I., Caciagli, P., Guarrera, G.M., and Nollo, G. (2006). Sterility and microbiological assessment of reused single-use cardiac electrophysiology catheters. Infect. Control. Hosp. Epidemiol. *27*, 1385–1392.

Traba, C., Chen, L., and Liang, J.F. (2013). Low power gas discharge plasma mediated inactivation and removal of biofilms formed on biomaterials. Curr. Appl. Phys. *13* (Suppl. 1), S12–S18.

Vandervoort, K.G., and Brelles-Marino G. (2014), Plasma-mediated inactivation of *Pseudomonas aeruginosa* biofilm grown on borosilicate surfaces under continuous culture system. PLoS One *10*, e108512.dol; 10 1371/journal pone 0108512

Rowan, N.J., Espie, S., Harrower, J., Anderson, J.G., Marsill, L., and MacGregor, S.J. (2007). Pulsed-plasma gas-discharge inactivation of microbial pathogens in chilled poultry wash water. J. Food Prot. *70*, 2805–2810.

Zelaya, A.J., Stough, G., Rad, N., Vandervoort, K., and Brelles-Marino, G. (2010). *Pseudomonas aeruginosa* biofilm inactivation: Decreased cell culturability, Adhesiveness to surfaces, and biofilm rhickness upon high-pressure nonthermal plasma treatment. IEEE Trans. Plasma Sci. IEEE Nucl. Plasma Sci. Soc. *38*, 3398–3403.

Index

Current Books of Interest

www.ingramcontent.com/pod-product-compliance
Ingram Content Group UK Ltd.
Pitfield, Milton Keynes, MK11 3LW, UK
UKHW022034041026
15591UKWH00012B/675